Air Carrier MRO Handbook

Air Carrier MRO Handbook

Maintenance, Repair, and Overhaul

Jack Hessburg

McGraw-Hill

New York San Francisco Washington, D.C. Auckland Bogotá
Caracas Lisbon London Madrid Mexico City Milan
Montreal New Delhi San Juan Singapore
Sydney Tokyo Toronto

Library of Congress Cataloging-in-Publication Data

Hessburg, Jack.
Air carrier MRO handbook : maintenance, repair, and overhaul / Jack Hessburg.
p. cm.
Includes index.
ISBN 0-07-136133-2
1. Airplanes—Maintenance and repair. 2. Airlines—United States. I. Title.
TL671.9H47 2000
629.134'6—dc21 00-050043
CIP

McGraw-Hill

A Division of The ***McGraw-Hill*** *Companies*

1 2 3 4 5 6 7 8 9 0 DOC/DOC 0 5 4 3 2 1 0

P/N 137242-3
Part of
ISBN 0-07-136133-2

The sponsoring editor for this book was Shelley Ingram Carr, the editing supervisor was Carol Levine, and the production supervisor was Sherri Souffrance. It was set in Garamond per the Gen1AV1 design by Joanne Morbit of McGraw-Hill's Hightstown, N.J., Professional Book Group composition unit.

Printed and bound by R. R. Donnelley & Sons Company.

This book is printed on recycled, acid-free paper containing a minimum of 50% recycled, de-inked fiber.

Dedication

Dedicated to the airline maintenance community—engineers, mechanics, and regulators. You may be featured neither in romantic flying novels nor TV movies but you make commercial air transportation the safest form of travel. Without your professional integrity, skill, and love of the business the airplane doesn't fly. Oh, and a mechanic's signature in the logbook. Thanks F.O.

Contents

Foreword

Jack Hessburg has a background well suited to provide a valuable reference book on air carrier maintenance operations. His long career at Boeing was centered around worldwide customer airline support. His eventual assignment as the principal maintenance representative on the B-777 design team helped Boeing provide a customer-friendly, easy-to-maintain, and extremely reliable airplane.

Jack Hessburg knows his subject, and this book can help anyone to understand the complex issues of U.S. air carrier operations.

Gordon Bethune
Chairman, Chief Executive Officer
Continental Airlines, Inc.

Preface

This is neither a management nor legal text. The world has enough of both. This book explains United States air carrier maintenance operations. Air carrier maintenance today evolved from commercial maintenance practices developed during the 1960s when we began to adopt on-condition maintenance practices. This evolved into the reliability-centered maintenance concepts of today. I am not inventing anything new. There are hundreds of individual references published by various agencies, Advisory Circulars, handbooks, orders, policy statements, annexes, legal opinions, engineering standards, reliability texts, etc. Unfortunately, each addresses only one subject in excruciating detail or regulatory doublespeak. I wrote this book because no single text explains the concepts and the system.

Much of what I discuss in this book is in reply to questions asked of me during the 20 plus years I taught an after-hours air carrier operations course at the Boeing Company. My intended audience is engineering and mechanic technology students, new entrant individuals at the airlines and manufacturers, and practicing mechanics.

The book explains the technical and regulatory characteristis of air carrier maintenance. It is a survey of the topic, designed to provide readers the means to construct a mental model of the subject upon which they may build detailed proficiency. Consequently, I do not delve into the detailed legalistic minutia of each regulation. Rather I present their effective character. Readers with a particular interest in a given regulation or maintenance practice will look it up to obtain the details. My apologies to the pedants among you.

Much of what is presented is typical of the industry. However, each airline has unique structure and practices to meet its own business and regulatory requirements. Nevertheless, each airline (foreign or U.S. domestic) will include the elements discussed in the book.

Don't be confused by the differences in structure or terminology you encounter.

The benefit of contemporary maintenance technology is that we can achieve the true intent of maintenance, the management of failure. Don't be fooled by those who say that airplanes and maintenance programs are highly technical and thus not readily understood by nonpractitioners of the discipline. Usually these people don't understand it. Airplanes have more detail and include design features different from another time. But they are just flying machines and they still break. Know your basics, and no matter the subtle complicated airplane designs or variation of maintenance practices you encounter you will understand that "airplanes is airplanes" and "fixing is fixing."Airlines are still a fun business. Remember to have fun!

The book is divided into four sections you need to know: contemporary issues; the law and regulations governing the aviation system; airworthiness; and the maintenance system that is the basis of assuring airworthiness.

Jack Hessburg

1

Contemporary Issues in MRO

No Fault Found Measurement—What's It Good for?

Thomas H. Carroll III

Everyone is looking for a way to benchmark the effectiveness of various processes and organizations. One of the most scrutinized areas of an airline is line maintenance effectiveness. Someone, somewhere, decided the best measure would be the "no fault found" (NFF) rate of components removed during the resolution of a system problem. And it sounds logical—if the maintenance process were efficient, they would remove only the components that actually have a fault.

At "Utopia Airlines," the sky is always blue and the flights depart on time, the mechanics work on perfect problems utilizing perfect troubleshooting, and the repair facility performs the perfect bench test and repair of the faulty components. Life is simple and straightforward, so NFF measurement is perfect for determining the effectiveness of line maintenance.

However, at "Real Life Airlines," the sky isn't always blue and the flights are often late, the problems and troubleshooting are very much imperfect, and the repair facility's bench test and repair processes are lacking. Life is very complicated at this and every other airline in the world, and using NFF measurement to determine line maintenance effectiveness is the most erroneous metric of all. How can that be?

Imperfect problems

The problems at "Real Life Airlines" are far from perfect, in the sense that many times the system fault occurs only during a specific flight regime, under dynamic flight conditions. Because of this, when the mechanic is performing system troubleshooting on the ground, the fault remains elusive. Now what? If everything checks normally under these static conditions, should the mechanic simply sign it off as "unable to duplicate"? How many repeat complaints will it take before some kind of action is warranted? Then when action is taken, what kind of troubleshooting tree should they follow when there is no problem present?

In dealing with these imperfect problems, typically the only course of action is to replace the most likely part and get on with it. Using this method, there are often a number of NFF shop findings before the problem is resolved.

Imperfect troubleshooting

As with anything involving human factors, everybody has good troubleshooting days and bad ones. Additionally, the short turnaround times during the operating day do not allow for in-depth analysis before taking action to resolve an elusive system problem. It gets down to a business decision whether to delay this and successive flights in order to go through detailed troubleshooting (which will probably be unsuccessful anyway), or take a chance of incurring an NFF charge by replacing the most likely part. In almost every case, it is more cost-effective to pay the NFF.

Then there are those systems with "built-in test" assistance, which is often erroneous and/or misleading. The needle of truth may be buried in a haystack of other messages. These factors also add to the NFF rate.

Imperfect bench check

The biggest contributor to NFF is the component bench check. Rather than being the precise measurement of component functionality, the bench check does not mimic the component's complete operation in the system. It is nearly impossible to check everything it does in every situation while interfacing with every associated system part. So if the test approaches 90 percent of system operational functions, it will leave a 10 percent margin of functional failures that can occur without any means to identify on the bench.

Furthermore, the conditions of the bench test are very much different than the operational environment of the component. I'm reminded of a control panel that would not exhibit failure until it was actually fastened into a rack. The standard bench test did not direct installation of the component as it was fitted into the aircraft, so it became a rogue unit. A rogue unit is a component that repeatedly experiences short service periods, exhibiting the same system fault each time, whose failure cannot be detected by standard bench test or overhaul procedures, and whose replacement on the aircraft resolves the system fault. In essence, it is a defective unit that never gets fixed, no matter what is done to identify and resolve the problem.

The most insidious thing about rogue units is that there is a "natural selection" phenomenon that ensures they will displace the serviceable spares. It's pretty much like a Darwin thing, but instead of survival of the fittest, it's more like survival of the worst. The gathering place for these rogue units is the spare pool—after all, being defective, they won't stay on the aircraft, so where else can they go? Furthermore, if they are not recognized and resolved, the spare pool will become more and more polluted until there are no spares that are actually serviceable.

When rogue units develop in an airline's inventory, the NFF rate skyrockets. The way it happens follows: During the troubleshooting of an elusive problem, the most likely part is installed, and it is a rogue unit pulled from the spare pool. The system complaint continues because a defective part has been introduced. Logical troubleshooting trees don't direct replacement of the same part again, so the other related components are replaced systematically, to no avail. All these related parts score NFF in the shop, because there was nothing wrong with them in the first place. When the rogue unit is finally replaced with a "good" spare, it goes to the repair facility to score NFF, then back to stock waiting to wreak havoc again.

Let's make this nightmare worse: Think of what happens if, during the course of replacing the related parts, another rogue unit was installed—now there are two bad parts causing system problems (the originally installed rogue unit plus this new one). This troubleshooting nightmare will go on for a very long time, because it defies logic that there could be two defective parts causing a single problem.

Again, this goes back to the inability of the shop to perform a bench check that mimics the complete operating functions and environmental characteristics of the component when it is installed. It is not an intentional situation, but a fact of life.

A few more facts of life: Every component has the opportunity to become a rogue unit. I've even seen rogue antennas! Also, there will always be rogue units, because over the course of time new and bizarre failures will occur in that unchecked region of the test.

The key to recognizing rogue units is to track each component by serial number, showing the date installed and removed, the aircraft on which it was installed, the flight hours and cycles accrued, hours since overhaul, and a reason for removal code. The removal codes should be very simple—if hundreds of codes are offered, human factors will reduce it to only a handful anyway.

Additionally, the aircraft history needs to be archived and easily retrieved in order to determine the exact system effect of the component's failure and whether its replacement resolved the problem. Finally, it would be good to have the shop history in a database in order to determine the work performed during previous shop visits.

If you don't track by serial number, disregard this whole section. You will never be able to identify rogue units, or determine component work scope, aircraft maintenance program shortcomings, or whether an aircraft or component reliability modification is really needed. Your best course of action would be to send everybody to brainwashing sessions so they believe they work at "Utopia Airlines," and the OEMs really know everything about their components and what is best for the airline operator.

What's it good for?

In a quick review, there are circumstances driving NFF that are completely out of the control of the line maintenance process; there are times when generating an NFF is an excellent business decision and the component bench check is not complete. All facts considered, it is obviously a poor measurement of line maintenance effectiveness. This is not to say it is completely worthless. A high NFF rate should prompt an investigation to see what is driving it—whether it is line, the shop, or both.

How can you tell?

The only good measurement of the effectiveness of both line maintenance and the shop is the aircraft system itself. When a part is replaced that fixes the problem on the aircraft, the evaluations can begin. The line maintenance process can be evaluated by how long it took to arrive at the resolution, as well as the shop's ability to identify that component's failure and repair it.

When the line process appears lacking, the next step is to determine whether it is because the problem was nearly impossible to troubleshoot, the built-in test lied, more analytic training is needed, or the system is just one of those prone to business decisions to keep the aircraft moving.

If the shop arrives at NFF for a component that finally fixed the system problem, then special techniques need to be employed to duplicate the problem. Obviously the test is lacking, so new experimentation must be utilized to duplicate the system operation and environment.

A word of caution regarding this experiment: If the unit doesn't get hot or freeze in normal operation, don't expose it to those extremes. If it doesn't vibrate like a paint shaker, don't do it. If it is failing in the normal parameters of its operation, duplicate those. Any extremes would be considered performance of highly accelerated life testing, which is intended to develop premature failure of subassemblies within the component. When using these desperate measures you'll find a problem, but the rogue failure will probably still be there.

The bottom line

The bottom line is that there is no one number that tells the whole story. NFF, mean time between unscheduled removals (MTBUR), mean time between failures (MTBF), or any averaged value will most often mislead rather than be an absolute guide. The only way to get a grip at "Real Life Airlines" is to gather aircraft history, track component installations and removals as previously described, and develop methodology to link these data during analysis.

You probably won't be able to change the name to "Utopia Airlines," but there will be more intelligence driving the decisions to control cost and improve processes, both on the line and in the shop. Hey, maybe the sky is turning a little bluer after all.

Thomas H. Carroll III

Tom Carroll has held the position of Manager of Reliability Engineering at US Airways since 1994. He joined US Airways in 1980, after 7 years as an avionics specialist in the U.S. Air Force, working both the line and shop disciplines. His first 4 years at US Airways were spent in the instrument shop, where he became proficient in the repair of more than 100 different rotables and repairables. In 1984, he became a technical foreman in the Avionics Technical Services Department, responsible for the operational performance of all avionics systems on US Airways' nine fleet types. Throughout the 9 years in that capacity, it became very apparent that component reliability was a large factor in contributing to chronic aircraft avionics problems, so he transferred to a newly created position as a component reliability engineer, where he was tasked with building a component reliability program for all rotables. It took about a year to get the program up and running, after which he was promoted to the position he currently holds. This enabled him to overhaul the entire airline reliability program. Since then, he has developed a comprehensive and successful program to control the development and effect of rogue components, in addition to new methodology to monitor overall component reliability.

He has spoken on the subject of component reliability and performance measurement at several industry meetings and conferences, as well as conducted training classes at OEM component repair facilities and aircraft manufacturing sites.

Uneasy Seatmates

Perry Flint

This paper had its origins in a speech presented to the IATA Aircraft Maintenance 1998 Conference and Exhibition, Dubai, UAE, November 3, 1998. When first I addressed this subject, I was not prepared for the positive response my brief remarks received and for the continued interest as they were disseminated across the airline community. I don't flatter myself that it was the originality of my thoughts, or the way in which they were expressed, that sparked such enthusiasm. Quite the opposite. Obviously, I struck a chord that resonates with aviation professionals around the world. Unfortunately, in the short period since my remarks were delivered, I have seen little that gives me hope that the trends I discussed are being reversed.

Safety and outside forces

Among these, probably the most disturbing is that the aviation safety agenda too often is being shaped by outside forces and directed increasingly into what I would argue are areas peripheral to the long-term goals of reducing the accident rate and saving lives. But I am also worried that the aviation profession and society at large no longer even appear to speak the same language when it comes to safety and understanding and managing risk.

I have no wish to dispute with those of you who believe you've seen it all before. I know that, in the past, safety agendas have been manipulated by individuals and groups. But in my 17 years of covering the industry, I have never experienced a period in which the influence of political forces was as overt and immediate in the safety process. Never before has "safety" been so broadly defined, to encompass items having little relevance to what, a decade ago, we would have considered aviation safety.

I use the word "political" broadly, to include not only governmental forces, but also the wide range of public and not-so-public interest groups that claim a right to define safety issues and to speak to them.

These can also be unions and manufacturers. Even airlines can decide to promote an issue as safety related, for reasons of competitive necessity or in order to forestall pending government regulation.

Furthermore, we live in a world in which emotions seem always to be just below the surface of things. We are constantly being called upon to react, to feel, to respond to external stimuli such as the horrific imagery of a plane crash or the heart-breaking stories of victims' families. The now common practice of bringing relatives to a crash site and keeping them there for weeks at a time, when they should be at home finding comfort in friends and family, is a boon for television networks and accident lawyers. But it also has helped to create an atmosphere in which sober reflection is simply impossible and the industry must respond helter-skelter to each new crisis as ever more subjects come under the media microscope.

Verdict first, evidence later

I would not deny any group a right to be heard. But we seem to have moved beyond reasoned debate and discussion, and to paraphrase *Alice in Wonderland*, have adopted a "verdict first, evidence later" approach to safety regulation. Those who attempt to slow down the process, to bring balance, often find themselves assailed as being "antisafety," or guilty of "putting profits ahead of safety." Rarely have governments moved so swiftly to address perceived safety inadequacies, even in instances where no direct harm can be shown to have occurred.

Swissair 111

A classic example occurred in the immediate aftermath of the crash of Swissair 111. Only a few weeks after the accident and before any finding of probable cause, the world aviation community was stunned to hear that the Federal Aviation Administration had determined that insulation blankets in almost all of the world's then 12,000 plus passenger jets (the number is now closer to 14,000) would have to be replaced, because installed materials had been deemed to provide insufficient resistance to the spread of fire. It may be that insulation ultimately is implicated as a contributing cause in the crash of the aircraft. But let's consider that fires on board aircraft are extraordinarily rare and there is no evidence of an injury or fatality attributable to an insulation blanket fire.

On the basis of the evidence of new testing that insulation blankets are not nearly as fire resistant as was supposed, I don't think the airline industry disputes that something needs to be done. But FAA's decision to release this information in a statement to the news media before the establishment of a rule-making process to review the issue, before any kind of cost/benefit analysis, which is required under U.S. law, I might add, and before even the drafting of new standards for insulation, represents a way of doing business that does not bode well for an industry that must already pay for items like Enhanced Ground Proximity Warning Systems, better pilot training, and improved navigation and landing aids that will have a direct, positive bearing on reducing the accident rate.

ValuJet 592

How have we come to this state of affairs? I believe that the roots of the present condition lie in the crash of ValuJet 592 in May 1996 and the vast media-directed outcry against the FAA that followed. The industry has long known that that accident occurred because an employee of a third-party maintenance company put a mislabeled box of oxygen generators into the belly of that airplane. This was a clear violation of hazardous materials shipping procedures and rules. Had that person complied with the rules, the tragedy need never have happened.

Outside the industry however, the belief was and probably still is widespread, that the accident occurred primarily because FAA failed to adequately supervise ValuJet's operations. This is owing in large part to a media-directed campaign on behalf of a former DOT employee who had published a book highly critical of the agency.

I believe the resulting frenzy of negative publicity had a crushing impact on the agency's sense of self-confidence and has led it to fundamentally reevaluate its approach to regulation. Quite simply, FAA has determined never again to be made a victim of media-manufactured outrage over perceived safety failings. And we have seen since that accident how it has interpreted the word "proactive" to introduce vast new regulatory programs at immense cost and on extremely short notice. All are beneficial, but it is by no means clear that they reflect the best use of available resources.

Consider that the cargo hold smoke detection and fire suppression rules mandated after the ValuJet crash have cost the industry a lot of time and money, but will have little effect on safety because such accidents are exceedingly rare. This, of course, pales in comparison to the massive cost associated with the antiterrorism security measures imposed by the United States after the crash of TWA 800, an accident that we now accept had nothing whatsoever to do with terrorism.

Trade unions

The ValuJet accident had another impact on the safety agenda. It gave new blood to an organized labor movement that has never shied from using safety as a flag behind which to advance its interests.

Let's begin with an irony: For most of the world, the role of trade unionism in industry either is stagnant or in decline. The only exception of note to this general rule, is the airline industry, where organized labor continues to attract new members while wielding enormous clout in the court of public opinion and at the bargaining table.

In defense of trade unions, I recall a remark by a highly-respected airline industry leader who once told me, "Companies that get unions usually deserve them." So if I take unions to task, I do so with the understanding that they are on the premises because employees no longer felt that they could get a fair deal without them. Furthermore, airline unions, in particular the pilots' associations, have and continue to make outstanding contributions to aviation safety.

Nevertheless, for the purposes of our discussion, the relevant point is their growing ability to refocus the safety agenda into areas that have not traditionally been considered safety related or to try to win through government regulation that which cannot be achieved at the negotiating table. Because ValuJet was a nonunion airline, and because of the circumstances surrounding the accident, its nonunion status became a rally point for unions, which did their best to implant in the public mind the idea that nonunion airlines are by definition unsafe airlines. Recently, the unions tried to use the legislative process to limit the ability of U.S. airlines to access non-U.S. repair stations, except in emergencies. In a global industry like aviation, a protectionist stance is unacceptable. It would have had a negative impact on many fine international airlines. But it is painted as a safety issue by the unions.

This is by no means a tactic that is confined to the United States of America. I still have a pamphlet from a group called the International

Transport Worker's Federation warning that the 1997–1999 "Asian crisis threaten[ed] aviation safety." Why? Primarily because nonunion airlines are cutting corners to save money and offer lower fares, according to the pamphlet. Of course, the spate of air disasters in Asia largely have involved airlines with a long history of organized labor and most occurred before the Asian economies collapsed. The ITF also repeats the old canard that competition "creates pressures which can seriously undermine safety," an argument that cannot be defended by comparing the safety records of those economically regulated airlines in Asia with the deregulated carriers in the United States.

Safety issues

Over the past few years, other items have been pushed to the fore as "safety" issues by unions. Generally these efforts can be exposed for what they are as long as they remain parochial enough to escape the attention of the mainstream news media. Thus, when FAA proposed a massive rewrite of airline pilot flight duty/time regulations that would have added millions of dollars in new costs, without improving safety, the issue never caught on in the media. As a result, airlines were able to put enough pressure on regulators to withdraw the rule quietly.

But as lifestyle issues, such as tobacco smoking, enter the mainstream debate, they are finding their way into labor negotiations as safety issues, which is how they are presented to the media at large. The debate on smoking, for example, inevitably becomes a discussion of passive smoke, and from there turns into a discussion of flight attendants' exposure to it as a serious health risk and gradually becomes transformed into a much broader health and safety issue for the traveling public. Contributing to this problem is the popular perception that cabin air quality is generally poor and that cabin ventilation systems can cause a buildup of contaminants such as cigarette smoke and contribute to the spread of disease and other undesirable effects.

Although numerous scientific studies have shown these concerns to be unfounded, they nevertheless are accepted as fact by many. A fair deal of cynicism surrounds the discussion. Experts with whom I spoke say that the cabin air-quality issue only surfaces around the time that flight attendants are due to negotiate a new contract.

But I do not doubt that in the dark recesses of the FAA or at the Environmental Protection Agency, bureaucrats are even today toiling away in an effort to regulate cabin air quality. When these regulations emerge, they will be presented as a safety issue, because they ostensibly address matters of public health. There are many similar issues floating around today. In each case the means by which they enter the mainstream of the safety debate is the same: a "problem" is identified. As it receives widespread dissemination in the broadcast news media, and now via the Internet, it acquires a veneer of legitimacy and urgency that propels it to the top of the agenda. Government, eager to be seen as doing something, responds with a new regulation.

The industry also is partly to blame for this. Cabin baggage, for example, is not primarily a safety issue; it is an operational issue. As passengers bring more and more luggage into the cabin the ability to load and unload the aircraft promptly is compromised, hurting schedule integrity. But because airlines are unwilling to act as their own police, they have allowed cabin baggage to be painted as a safety issue and have invited the federal government to regulate the practice. The law of unintended consequences virtually guarantees that any regulation will create far more problems than it solves. In fact, it already has. As part of its evolving safety agenda, the FAA has now assigned a full-time cabin safety inspector to every major U.S. airline. I look forward to the first fine assessed against an airline for having a bag protrude from under a seat.

You're probably wondering how an agency that is already hard pressed to carry out its existing regulatory and air traffic responsibilities can afford to assign full-time inspectors to issues like in-flight baggage stowage, child safety seats, and the like. The answer is that public opinion has been shaped to believe that these are important airline safety issues.

Assessing risk

Today, an added dimension to this problem exists that gives me great concern. Our society's ability to understand and accept the concepts of risk and risk measurement is rapidly disappearing. Indeed, we are faced with a paradox: With each new advance in science, technology, and medicine that improves the overall quality of our lives and increases our average life expectancy, comes a corresponding unwillingness to accept the ever-shrinking amounts of actual risk in our personal lives. Consider the following examples.

The value of life

This affects aviation safety in a couple of ways. The first is that we cannot discuss safety improvements logically and rationally. It has become very difficult, for example, for FAA to publicly defend the key equation in any safety argument, which is the value of an individual's life in dollars, measured against the overall cost to society of each improvement in safety. This figure is currently $2.7 million but I fear that we are very close to the point where the economic value of a human life will be deemed to be infinite—in practical terms we already may have taken this step—and this industry will be expected to expend all available resources in order to improve, by some imperceptible amount, the already enormous margin of safety in the airline industry, regardless of the overall consequences to society.

Child safety seats

Now, the corollary to this trend is that we have become statistical illiterates. Thus we are incapable of grasping that some risks are very, very real, whereas others have an insignificant impact on our lives, and that there are very relevant trade-offs between them. A case in point is the FAA's recent announcement that it intends to require that all children under the age of 2 travel in child safety seats on board aircraft. From a political and public relations standpoint, in the post–ValuJet era, such a rule may be a "no-brainer." Statistically, however, it is indefensible. Data show that, at most, two lives over the past 20 years might have been saved had the children been in safety seats during airplane accidents. Against these two lives must be weighed the thousands of children who die each year in car accidents. How many more will be lost because families choose to drive rather than accept the cost of an additional airline seat?

Heart defibrillators

Another example is the carriage of heart defibrillators on board aircraft. From the amount of attention paid to these devices in the mass news media, one could only come away with the impression that thousands of people are dropping dead every year from heart attacks and other coronary events while en route somewhere. Now, I'm not saying that having defibrillators on board aircraft is a bad idea. But there is no way that the investment can be justified in terms of the actual number of annual in-flight emergencies that require such devices. Nevertheless, the FAA will require that U.S. airlines—

and that means inevitably their non-U.S. counterparts that operate to the United States and eventually the airlines of many other nations, invest in these devices, which cost around $3,500 apiece. The FAA's decision is based on a year-long study, in which defibrillators saved the lives of four passengers in 17 instances in which they were used. FAA has estimated the cost to the industry for the devices and training at $138 million over 10 years. Of course, as medical technology advances, so will defibrillators and I'm sure airlines will be expected to stock the latest, most expensive devices on their aircraft.

How did defibrillators on aircraft come to be seen as an aviation safety issue? Probably in part owing to aggressive marketing on the part of manufacturers of medical equipment who see a relatively large, untapped market for their devices, particularly as our population ages and becomes statistically more vulnerable to coronary events. People no longer "slow down," as they get older. Partly it is because of genuine concern with airlines that operate a significant number of extremely long flights over water and remote areas, where emergency landings are impossible and adequate medical facilities few and far between. And, of course, as these airlines introduce the devices it becomes paramount for their competitors to do the same.

At least defibrillators represent a response to a real problem. But what are we to make of the U.S. Department of Transportation's fortunately unsuccessful effort to regulate the serving of peanuts on board commercial aircraft flights operated to, from, and within the United States? This, too, was advanced under the banner of safety: to wit, to protect from harm those persons who suffer from peanut allergies.

Was there a public outcry in favor of regulation? No. Are there any studies to support the position that the serving of peanuts represents a serious public health issue that needs to be addressed at the highest level of government? Of course not. But industries—and regulators—have become increasingly vulnerable to pressures that can be brought by tiny, well-organized special interest groups able to seize on a seemingly worthwhile health goal and generate favorable publicity for their cause. The Internet has proved a godsend to these groups, which now are able to organize and disseminate their views across the globe at virtually no cost to themselves.

This, then, is the shape of the new safety agenda. It is driven by a combination of little-known but media savvy interest groups, and the immense pressure on broadcast news media to generate higher profit margins, which can only be achieved by showing ever more shocking, more outlandish, and more entertaining stories.

Environmental issues

Of course, even issues like peanut bans and defibrillators are relatively benign in the vast scheme of things. But as aviation safety is redefined to include environmental issues—which are now viewed as public health matters—we are seeing more intrusions that actually have the potential to negatively affect safety.

At Amsterdam's Schiphol Airport, for example, pilots have been asked to take off and land from runways selected for environmental considerations, not flight safety. Now, what we're talking about here is noise pollution, so the risk of actual harm to humans or the environment is not quantifiable. But we do know that around 63 percent of all aircraft accidents occur during takeoff, final approach, or landing. Yet regulators potentially add to the workload at this critical time.

Broader picture

Obviously, this is one of the more outrageous examples of how the safety agenda have been reshaped in a manner that is degrading safety as we understand it. The broader picture that emerges, however, is one of a schism between what aviation safety professionals view as the key issues to be addressed and the direction in which the safety agenda are often being pushed. Although it might seem possible to proceed on two unequal tracks to arrive at the same goal, the reality is that resources are limited. Even if every dollar of airline revenue was dedicated to the cause of safety, it is still not an infinite pot of money. Thus, the industry's ability to tackle the two biggest problems, CFIT (controlled flight into terrain) and loss-of-control accidents, is compromised so that serious resources must be allocated to issues like carry-on baggage, disruptive passengers, cabin air quality, child safety seats, heart defibrillators, superfluous antiterrorism measures, and environmental health. And I've not even touched on 16G seats!

Identifying a problem is not the same thing as being able to solve it. We have all seen "the graph" showing that as traffic grows, so will the absolute number of accidents, unless the accident rate can be reduced. But, of course, long before the industry sinks to the level of one fatal accident per week, governments will impose such restrictions on operations that the rate of traffic growth will be reduced to a fraction of what it is today.

The industry has embarked on several projects—the Commercial Aviation Safety Team and the Aviation Safety Alliance—that it hopes will be able to put it back on course in terms of focusing on the critical safety issues. In the meantime, my plea is that everyone with a commitment to improving aviation safety continue to stay focused on the real safety issues. The choice is clear: If we lose sight of what really matters in aviation safety, we will be as guilty as those for whom principle long ago was sacrificed on the altar of self-interest.

(The opinions expressed here reflect those of the author.)

Perry Flint

Perry Flint has been reporting and writing about the airline and civil aircraft manufacturing industries for nearly 17 years. He currently is executive editor of *Air Transport World Magazine*. Mr. Flint has written extensively about all aspects of the world airline and commercial aircraft manufacturing industries, including aeropolitical and regulatory, legislative, commercial, financial, and technical.

He was short-listed for writing honors by The Royal Aeronautical Society in 1996, 1997, and 1998 and was named an Aerospace Journalist of the Year by the society in 1996. He was honored by the Aviation Space Writers Association in 1993 for his reporting on the changing relationships among aircraft makers and their suppliers.

Mr. Flint is a recognized authority on the airline industry and a regular speaker at industry forums. He also has appeared on CNN and MSNBC as well as numerous radio news programs to discuss financial, regulatory, and safety issues related to the industry.

Mr. Flint began his aviation journalism career as associate editor of *Commuter/Regional Airline News* in 1984. He subsequently became editor of *Airline Financial News*. He joined the staff of *Air Transport World* in 1987 as finance editor. He was named executive editor in 1992.

Mr. Flint received B.A. degrees in history and literature from The American University in Washington, D.C., in 1980.

The State of Aviation Technical Education

Gary Eiff

Over the last 50 years, aircraft have evolved from straightforward mechanical designs into highly complex machines with sophisticated dependent and interactive systems involving the most advanced technologies. Considering the neoteric technology used in today's aircraft and the rapid expansion of the industry, it is prudent to consider the question of the adequacy of today's technical education in providing the knowledge, skills, and abilities necessary to maintain these complex systems.

Need for change

Change begets the need for change. Nowhere is this truer than in the aviation technological workplace. The first concerted attempt to assess the state and needs of aviation technical education came in the late 1960s as the industry evolved from internal combustion engines to jets. Industry leaders questioned whether technical education curriculums could adequately prepare entry-level technicians with the fundamental knowledge, skills, and abilities to effectively perform their duties. This led to an extensive review by Dr. David Allen, in his 1966 study, of the knowledge and skills required for new-generation aircraft and a restructuring of the curriculum requirements for approved maintenance training institutions. In his Phase III report: A National Study of the Aviation Mechanic's Occupation, Dr. Allen stated:

> *The aviation mechanic plays a vital role in the air transport industry of our nation. The nature of his occupation requires that both initial training and subsequent in-service training provide him with skills and technical knowledge necessary to perform "return-to-service" work with the highest precision and efficiency possible.*
>
> *Technical Advances within the aviation industry are occurring at an extremely rapid rate. These advances have created*

> *a need for additional aviation mechanics to maintain the sophisticated systems of modern aircraft. These technological advances have also made it necessary to update the instructional program currently being used in most aviation maintenance technical schools.*

These two paragraphs, although written about problems facing the industry in the late 1950s, capture numerous issues central to the problems confronting the aviation industry today. It is well documented that the current aviation technical workforce is populated by largely older workers who will soon retire. These seasoned technicians represent a brain and skill trust that will have to be replaced soon by technicians entering the workforce. Coupled with the current rapid expansion of the airline industry, the need for highly qualified technicians conversant with the advanced technology of today's aircraft will reach critical proportions in the near term.

Aviation Maintenance Technician rating

The dependence of current aircraft on advanced electronics and highly automated system designs as well as new construction technologies such as composite structures has proved current technical education curriculums to be hopelessly obsolete and woefully inadequate to meet this challenge. Forward thinking industry leaders recognized this fact early in the 1990s and attempted to resolve the issue by proposing a two-step licensing strategy for maintenance technicians. The proposal sought to strengthen maintenance school curriculum requirements through the addition of much needed subject matter in electronics, composite materials, and systems instruction. Under the proposal, the current two license systems of Airframe and Powerplant ratings would be replaced by a single Aviation Maintenance Technician (AMT) license. Individuals seeking employment in maintenance environments dealing with large transport category aircraft would be required to complete additional training and qualify for the Aviation Maintenance Technician—Transport category (AMT-T) rating. This initiative to close the gap between technical education and the advanced state-of-the-art aircraft and systems technology of today's aircraft was met with considerable opposition.

Many maintenance schools opposed the rule change for several reasons. Most schools believed that they would have to offer both the advanced large aircraft AMT-T training as well as the fundamental AMT curriculum in order to be competitive. Many schools reported that they would be unable to afford the additional equipment needed to support the new curriculum requirements. It is also an unfortunate fact that the industry has not been as supportive of schools as it could be when it comes to providing equipment, training aids, or course materials for educational purposes. In fairness to aviation companies, this was often because of a failure of the schools to work closely with industry and communicate their needs in a timely and realistic way. Many aviation companies and organizations are now beginning to furnish more equipment and educational support for schools as well as provide access to many industry training classes.

Technical schools also expressed concern that the new curriculum material would add considerable time requirements to their existing programs. Junior colleges with aviation technical programs reported that they would not be able to support the additional time requirements because of their educational charters. This was interpreted as meaning that they would be limited to providing only the basic AMT education, which would place them at a competitive disadvantage to other schools when it came to recruiting new students. Many proprietary schools believed that the additional time requirements would adversely impact their ability to attract students into their programs. Declining enrollments over the past two decades have made aviation technical schools sensitive to this issue. Rapid expansion of other, competing technology programs has made it increasingly difficult to attract potential students into aviation programs and away from programs such as computer technology, computer programming, technical graphics, and other career preparation programs, which offer more lucrative and prestigious employment opportunities.

Students once pursued aviation careers because of its adventurous image, a love of aircraft, and the technological prominence portrayed by the industry. In today's high-tech world, airplanes no longer capture the imagination of young people like they used to. Increasing security concerns have placed airports, hangars, and airplanes beyond the reach of inquiring young minds. Gone are the days that many of us remember when young people could walk around on the airport visiting aircraft, facilities, and aviation professionals. Limited access coupled with the industry's dismal record of promoting itself has resulted in fewer young people seeking careers in aviation. For many young people, the only portrayal of the industry they are exposed to is that represented in the media or the often negative image painted by aviation professionals themselves of their careers.

Curriculum requirements

In an industry that experiences rapid technological change, such as aviation, it is imperative that career preparation education keep pace with emerging technologies in order to adequately prepare entry-level technicians to take their place in industry. A close examination of other technology programs, such as computer educational programs of various types, reveals a fierce dedication to keeping pace with rapid changes in technology related to their fields of study. Why, then, do we in aviation technical education find ourselves encumbered with the baggage of pre-World War II technology in today's maintenance curriculums? There is no doubt that stagnation in the subject matter, knowledge, and skill requirements of the Federal Aviation Administration's required curriculum for aviation technical education is a primary force in the failure of schools to adjust to emerging technology within the industry. Little has been done to update the aviation technical curriculum requirements since the Allen study. It is not surprising, therefore, to find "dope and fabric" as a requirement rather than knowledge and skill requirements for newer composite materials. Aviation technical education curriculum requirements are long overdue for extensive revision.

There is little doubt that today's aviation technical education is significantly dated and woefully inadequate to meet maintenance requirements of today's advanced aircraft and system designs. Highly integrated electronic components communicating over various types of computer bus architectures are at the core of today's maintenance and troubleshooting procedures, yet we require only a basic understanding of electricity in our technical curriculum. Highly sophisticated carbon-fiber composite structures permeate current aircraft designs, yet "dope and fabric" still haunts our training requirements. If we are to successfully meet the challenges of mass retirements within our workforce and the rapid expansion of the airline transport industry, we will have to move aggressively to restructure our technical curriculum to reflect the state of technology in current aircraft and system designs and provide curricular flexibility to accommodate the technical advances of tomorrow's aircraft. This can be achieved only through a cooperative partnership of government, industry, and technical education institutions.

Gary Eiff

Dr. Gary Eiff is a professor in the Aviation Technology Department at Purdue University. His primary teaching responsibilities are with the school's FAA Part 147 Maintenance Technician Education program. He teaches undergraduate courses in basic electricity, avionics systems, flightline avionics troubleshooting, and advanced avionics systems. Dr. Eiff also teaches graduate-level courses in human factors, error management, and human error.

Dr. Eiff's research interests center around aviation human factors and organizational error management strategies. He is widely regarded as an expert in aviation maintenance and operational human factors and works on numerous applied research projects with aviation industry partners from all over the world. Dr. Eiff has written over 70 conference and journal articles and speaks at numerous international conferences and meetings throughout the year.

Dr. Eiff has over 34 years of aviation maintenance experience and holds an A&P mechanics license as well as a Repairman's Certification for Radio Class 1, 2, limited 3, and limited instrument. He is also a commercially rated pilot and flight instructor with over 4,800 hours of flight time. During his industry tenure, Dr. Eiff performed the duties of mechanic, avionics technician, supervisor, shop manager, general manager, and chief executive officer at various aviation companies.

Dr. Eiff received his master's and doctoral degrees in education from Southern Illinois University. He has held numerous offices in aviation organizations including the Association for Avionics Education, Aviation Technician Education Council, University Aviation Association, International Society of Aviation Maintenance Professionals, and Women in Aviation International.

Bogus Parts

Elizabeth A. Marchak

Along the streets and alleys near the Miami International Airport, there are hundreds of aviation repair stations small and large that do repairs, overhauls, and maintenance. Nestled among them are shops that do fake repairs or no repairs at all. Many of them have one thing in common: their parts, good and bogus, end up on the nation's airliners.

Gabriel Kish used to buy and sell bogus and stolen parts in Miami, he told the Senate Government Affairs Subcommittee on Government Operations May 24, 1995. Even though an FAA inspector would stop by his business, the inspector never figured out that Kish's shop lacked vital equipment to test and repair parts. The inspector was more interested in wooing his secretary, Kish said, than ensuring aviation safety.

Kish got an upgraded certificate that permitted him to make more repairs. It was delivered by a second FAA inspector who didn't inspect any equipment either, he said. "As before," Kish told Senator William J. Cohen, Maine Republican, who chaired the hearing, "my equipment was not operational or calibrated."

Like others who offered up unairworthy parts, Kish was cashing in on the industry's perpetual need for spare parts. In a transportation crossroads like Miami, the FAA can't begin to keep up with the lawbreakers. Kish, also known to the FBI as Gabor Kish, was nabbed in a large bogus parts ring called Operation Skycrook in 1993. He served a 2-year prison term.

At the hearing Kish testified along with one of the men who helped put him in jail. Harry Schaefer, special agent at the Transportation Department's Office of the Inspector General, said the OIG was devoting more and more resources to bogus parts each year. "We simply cannot investigate all the suspected unapproved parts cases that need to be worked with our limited resources," he said. "We need the help of the FAA on these cases," he said, making a plea for help from the FAA in all of its offices in all of its regions. "Over the past few years, the field inspectors have received mixed signals with respect to the FAA's SUP (suspected unapproved parts) program, ranging from red lights to yellow lights to green lights," he said. The inspectors "need a clear, steady green light from FAA headquarters," Schaefer said.

In the hearing room, oversized photos of red-tagged aviation parts, too worn or too damaged to be used in aviation, were on display. In several cases, government enforcement cases found that the parts are pulled from scrap piles and repainted, retagged, and resold to airlines flying passengers with yellow tags certifying them airworthy.

The problem of counterfeit, bogus, or unapproved parts is not new. The Civil Aeronautics Board warned of it in 1957. But it was not until after Kish's testimony and several unflattering stories about the FAA's reluctance to deal with the problem that the agency officially recognized the problem.

On Oct 12, 1995 the Federal Aviation Administration created the Office of Suspected Unapproved Parts, based at Washington-Dulles International Airport, to monitor the growing problem of bogus or unapproved parts. Margaret Gilligan, deputy associate administrator for regulation and certification, said the move was triggered by congressional inquiries and media scrutiny. The office, working with the OIG, the FBI, and investigators at other government agencies was supposed to collect, assess, and resolve the reports.

The announcement came the same day two New York men who supplied bogus oxygen systems and fire extinguishers to Air Force One and Air Force Two pleaded guilty to five felony charges. Their company, Air-Tec Services, Inc., lacked equipment to test and repair parts, even though it had an FAA certificate. Several companies were also affected by Air-Tec Services, including Continental, US Airways, AirBorne Express, Boeing Defense & Space Group, and Beech Aerospace.

Sting operations like the one that nabbed Kish and the two New York men made headlines. Scores of other bogus parts cases over the next 5 years didn't. Law enforcement officials realized they did not have the right law to prosecute those who knowingly bought and sold bogus parts.

In March the Aircraft Safety Act of 2000, was passed. "Additional investigations are under way with no sign of abatement," the law's preamble says. Under the new law, if a parts supplier says it meets

FAA standards and it doesn't, the person responsible will face much tougher criminal penalties if convicted.

Simple violations of the law carry maximum penalties of 10 years in jail and up to a $250,000 fine. If a conviction involves the quality of the part, the jail term may run as long as 15 years and the fine may double. If the part creates a "malfunction or failure that results in serious bodily injury," a 20-year prison term may be imposed along with a fine of $1 million. If the part causes a malfunction that results in death, the defendant may be sentenced to life in prison and a $1 million fine. For corporate offenders, fines as high as $20 million are written into the act.

The law also covers all electronic communication and paperwork associated with parts in the United States and abroad. Industry has estimated that as much as $2 billion in unapproved parts are now sitting on the shelves of parts distributors, airlines, and repair stations. That includes civilian, public, and military aircraft as well as aircraft and spacecraft parts.

Prosecutors and those in law enforcement don't expect they will have to wait too long to test the law. "The problems associated with nonconforming, defective and counterfeit aircraft parts are legion since a single domestic passenger airplane alone can contain as many as six million parts," the preamble says.

Elizabeth A. Marchak

Elizabeth A. Marchak specializes in computer-assisted reporting at *The* (Cleveland) *Plain Dealer* where she investigates aviation safety, including the spread of bogus or unapproved airplane parts, safety problems at ValuJet Airlines, and other topics. Based in Washington, Marchak makes extensive use of data and the Freedom of Information Act to examine systemic issues. She has taught numerous seminars to reporters on aviation issues, covering the FAA and using data to tell stories. Her work is profiled in the *Aviation Handbook* published in 2000 by investigative reporters and editors. Marchak is a recipient of the George Polk Award for National Reporting, the Cecil Brownlow Publication Award from the Flight Safety Foundation, and numerous other reporting awards.

Repair Station Report: Year 2000

Robert E. Doll

At the time of deregulation, nominally 1980, only a few of the major airlines had extensive in-house capability to handle all levels of aircraft and component work. Most airlines did not have the resources or the maintenance volume to invest in capital intensive repair (MRO) capabilities. For example, very few of the largest airlines had complete engine repair capability or complex avionics test and repair capability. This work was contracted out to third-party (independent) repair stations or to the larger airlines that had excess capacity to sell.

Deregulation spawned many new airlines searching for niches in which they could successfully operate without causing one of the majors to price them out of business. These start-up airline companies did not have the financial resources or the fleets that would justify investing in the equipment and infrastructure required to maintain their own equipment. Further, the equipment employed was, for the most part, surplus equipment acquired from the major airlines at a very low ownership cost offsetting higher maintenance costs.

The second dynamic that was developing in the aircraft aftermarket during the same time period was the growth of the cargo and package carriers. FedEx started to acquire used aircraft at a high rate in the 1970s. UPS had a growing package-by-air delivery business in the same time period and they began to acquire their own fleet. The cost of acquisition of these older aircraft types and the subsequent conversion to freighters by the various cargo and package companies provided economical lift that stimulated the whole airfreight market. At the same time, U.S. manufacturers were moving to "just-in-time" inventory schemes that required short (fast) distribution pipelines. The new airfreight carriers were there to satisfy this growing demand. Cargo fleets grew rapidly to support this business.

The package and cargo carriers never did invest in any substantial maintenance infrastructure. The combination of new entrant airlines, the rapid growth of several "national" airlines, and the airfreight carriers stimulated the MRO and passenger-to-freight conversion markets by employing older aircraft. The major airlines did not have nearly the capacity needed to provide MRO services to the industry resulting in the opportunity for the formation and growth of third-party repair stations. The growth of these companies and the expansion of new entrants has been rapid over the past 20 years.

Today, the MRO market is estimated at approximately $30 billion per year worldwide. The size of the market has stimulated many different types of service providers:

- Entrepreneurial businesses that specialize in segments of the aftermarket such as airframe maintenance and modification, component overhaul, engine overhaul, field installation teams, and line maintenance.
- Original equipment manufacturers (OEMs) that provide aftermarket support for their products and, in some cases, support for competitors products.
- Technical business units spun off from airline companies that operate as stand-alone profit centers providing MRO services for smaller airlines as well as the parent.
- Airline companies that have set up in-house business units to sell excess capacity.

Today, there are MRO providers all over the world attempting to meet the steadily growing demands for aftermarket services. The European carriers have been leaders in forming separate business units to handle their own maintenance and to sell maintenance services to other air carriers. Some have sold their repair facilities to private enterprises. Here in the United States, the major air carriers such as United, American, and Delta have set up business units in their organizations to handle the demand for their excess capacity. At the same time, these carriers will outsource significant amounts of work in the form of special projects that do not fit their routine workloads.

In 1996, the ValuJet accident brought a tremendous media and regulatory focus on third-party repair station operations. The public was led to believe that third-party work was a new, dangerous phenomenon even though the preponderance of work done at repair stations had been of the highest quality for many years. It was determined as a result of the investigation into ValuJet and their maintenance provider, Sabre Tech, that maintenance oversight by the FAA and by ValuJet was seriously lacking in that case. The credibility of the whole industry suffered as a result.

ValuJet violated the most basic tenant of airline operations: an airline can legally delegate authority to perform maintenance and engineering functions to a third party but they cannot delegate the responsibility for the performance of those maintenance functions to meet regulatory standards. For example, UPS has historically outsourced almost 100 percent of its maintenance work to third-party repair stations. UPS operates one of the larger fleets of aircraft in the world. They have set up an extensive oversight organization to assure that their maintenance work is done to their very demanding standards.

Since the time of the ValuJet accident, the FAA has tightened up their oversight of all repair stations, establishing new standards of quality within FAR 145. These new quality standards are designed to assure a quality environment at a repair station that is equivalent to the quality environment one would expect to find in an airline maintenance operation. The new regulations have caused repair stations to tighten up on their quality assurance functions and have generally increased operating costs. The new regulations had virtually no impact on repair stations that were operating to the higher standards either because of their own business philosophy or because they were pushed to those higher standards by customers like UPS or Southwest.

Today, repair stations in the United States are faced with the problem of finding and keeping qualified people, both technicians and managers, who can perform to the higher standards. The industry as a whole is faced with a daunting labor shortage. Repair stations find themselves developing technicians only to see them leave for the more attractive and higher paying jobs at an airline. The labor shortage will not be cured by traditional recruiting and public relations methods. The industry needs to find more productive ways to diagnose problems and repair airline equipment.

Repair stations are also being pushed to higher wage levels by the competition for technical workers from other industries, from the OEMs that are active in the aftermarket, and from European operators that are entering into business here in the United States. The competing industries and OEM-related MRO companies are offering pay scales that are up to $10 an hour more than the comparable repair station wages that currently prevail in the United States.

The competition from the OEMs and the Europeans will force domestic repair stations to operate to a higher level of productivity, reliability, and quality if they are to remain in business. The management challenge is to send equipment through a repair station in the shortest possible time, reducing the labor content wherever possible without sacrificing quality. Many repair station customers are operating on thin budgets and will not be able to absorb higher prices from the increased labor costs that will be passed on to them.

New repair stations are being established at many locations in the United States because of the availability of vacated military facilities. These facilities usually offer a very low occupancy cost through favorable lease terms in return for the creation of new jobs. Often maintenance and test equipment is available with the facility. The low overhead of these facilities may be enough to counter the increased wages that a new repair station will need to offer to man its facility. The successful new business will be the one that can provide its labor force with the most attractive wages, benefits, and stability, luring technicians away from competitors.

In summary, there is enough growth in the MRO market to sustain the expansion of the established repair stations and to support new entries into the aftermarket. Customers will insist on a new plateau of performance to account for the higher costs they will pay. The successful repair station in the future will need to work at less than arm's length with its customers. Significant improvements in maintenance planning will be required to create the productivity increases needed to cover increased costs of labor.

The opportunity for innovation in the management of MRO businesses is huge. The management team that can help its customers effectively plan their maintenance work and visits will find no bounds to their business success. Look forward to several years of turnover within the aftermarket before the new MRO business profile is defined.

Robert E. Doll

Robert E. Doll is president and CEO of Tech/Ops International. Bob has over 30 years of experience in the transportation and manufacturing industries as a line manager, an operations management expert, and a consultant. Prior to forming Tech/Ops International in 1993, Bob was a Maintenance Division vice president at United Airlines, a principal at KMPG Peat Marwick, a Maintenance Division vice president at Continental Airlines, a Maintenance Division director at Eastern Airlines, and a vice president and officer at K. W. Tunnell Consultants.

Bob has been recognized by the Air Transport Association (the 1993 "Nuts and Bolts" Award), the United States government (FAA Distinguished Service Award), the French government (DGAC Distinguished Service Award), the Flight Safety Foundation Laura Tabor Barbour Award, and *Aviation Week Magazine* Laurels Award for his work in furthering aircraft safety. He currently is chairman of the FAA RE&D Advisory Committee. Bob has a bachelor's degree in mechanical engineering from Pratt Institute and has taught statistics and operations research courses as an adjunct professor in the bachelor and master's degree programs at New York Institute of Technology and Nova University.

Some Thoughts on Safety

Jack Hessburg

Safety has been inherent to aviation since the work of Octave Chanute, Santos Dumont, the Wrights, and others. The untimely death of Lieutenant Thomas Selfridge and serious injury of Orville Wright in 1908 clearly affirmed that aviation is an unforgiving art. It is axiomatic that anything that places the airplane or its occupants in peril is unacceptable. Safety must manifest itself in more than sloganeering. It carries into the individual. Herewith are some of my thoughts.

As an industry we have collectively done an excellent safety job. We pride ourselves with astonishing statistics showing how safe flying is compared to other forms of transportation. We have training programs, wall charts, checklists, seminars, and new safety organizations. We all attend safety meetings and religiously repeat the cliché-laden mantras to each other. Stentorian pronouncements by presidential and congressional committees, all bent upon assuring the public of air carrier safety, have resulted in flashy safety initiatives responding to the public's demand "to do something" after every tragedy. We rehash the evidence and define attractive goals that are most likely unattainable and, if they are attained, will be statistically insignificant.

The truth is that we have done such a good job of assuring safety we have fairly well squeezed the turnip dry. We are not going to have significant improvements in safety statistics. We will merely hold our ground. Many of today's initiatives, although politically correct, will yield little toward improved statistics. We are expending valuable funds and talent chasing phantasms and solutions to many insignificant ill-defined problems.

If we are to really squeeze the last ounce of safety improvement from the turnip we must sweep out the corners of the safety envelope, not just seek chrome-plated, highly visible, costly projects. We must be introspective of our commitment of resources toward people problems, personal qualifications, ethics, and attitudes. We must reinvigorate the safety culture on a personal as well as an institutional basis.

Maintenance safety—a stepchild

Large sums are being spent to make sure we understand what causes pilots to make mistakes. Decision-making processes, accident precursors, pilot behaviors under stress, ergonomics of cockpits are all studied, researched, and discussed to the point of nausea. Don't misinterpret me. These efforts are valued. But they are given budget and focus to the exclusion of the remainder of the flight safety community and basic cultural issues.

The maintenance community must be a coequal at the safety table with the flight disciplines. Maintenance error injures and kills as easily as pilot error. Mechanics are people just like pilots. Are maintenance people any less stressed in their work? Are they less likely to be influenced by poor communication, distraction, poor problem-solving skills, and hostile environments? Of course not! The same influences and mental processes contributing to pilot error apply to mechanics. The only difference is the venue.

The man-machine interface in some instances is far more complex in the maintenance community. Human error, miscommunication, poor procedures, and pressured working environments contributed as much to the ValuJet accident as did unethical practice. Or consider the accident involving the failure of a compressor disk in a JT8 engine installed on a 727. Poor inspection equipment, inadequate lighting, manufacturing defects, inadequate inspection standards all contributed toward the failure to detect a flaw, which eventually resulted in death and injury.

Now statistics are a valuable tool. They are very useful for identifying immediate safety threats. They point to primary issues. But they are a misleading tool. Low accident percentages should not be neglected or dismissed as being insignificant. Any percentage is a beacon on the horizon and should not be summarily dismissed.

Maintenance is a relatively small percentage of the probable cause of air carrier accidents. Consider that these oft-quoted statistics address only air carrier–related accidents, not aviation as a whole. The truth is that the number of maintenance-caused accidents has increased since the early 1990s. But most appear to be associated with general aviation, not air carrier maintenance. Complacency leads us to believe that we are unaffected in the air carrier world. But maintenance is maintenance regardless of the setting. What is taking place in general aviation is a harbinger of the future for air carriers. Yet we don't seem to be expending substantive resources to isolate the cause of this increase within the commnity as a whole.

Further, consider that the air carrier statistics may be skewed. I don't know if the low air carrier percentages are because of superb design coupled with excellent maintenance practice or if they are a lack of comprehensive maintenance statistics. I think it is the latter. An informal review of NTSB accident reports dating back to the 1980s showed contributory maintenance factors toward an accident are poorly documented.

It appears we don't really know how much maintenance error contributes to accidents. We should!

The illusion of knowledge

> The greatest obstacle to discovering the shape of the earth, the continents and the oceans was not ignorance, but the illusion of knowledge.[1]

There is no place for superficiality in aviation. Yet presently we seem to be overwhelmed by it. The number of empty suits and talking hairdos in the business is increasing asymptotically. You must know your craft, not convey the impression that you do. Knowing your craft involves not just initial training but also continuing education throughout your career. Not doing so perpetrates a fraud. Meddling outside your discipline also constitutes a fraud.

The illusion of knowledge frequently leads to what I call interdisciplinary meddling. It is a recipe for a safety disaster somewhere down the line. It occurs when business managers presume to make flight operations decisions, or pilots make maintenance decisions or maintenance flight operations decisions, and so on.

[1]D. J. Borstin, former librarian of Congress and professor of history at the University of Chicago.

Nowhere is this more evident than in the pressures from Wall Street analysts or individuals whose position as business managers inures them with a myopic arrogance to presume expertise in technical issues. All too frequently under the guise of the "bottom line."

Missionary zeal is another form of the illusion of knowledge. Each of us has unique professional competence and credentials. We must not succumb to feelings that we must save each other from the error of our professional ways. However, we must demand of each other that we are competent to our individual disciplines. Demand that pilots know how to fly, that mechanics know how to fix things, that engineers know how to design, and that accountants know how to balance books.

Understand an "opponent's" discipline so as to intelligently debate issues. However, we must not presume to know the subtleties of the other's job. We must accept each other's professional expertise. Above all else, when conducting the debate it must take place among peers, we must not place each other on "trial."

Responsibility

The regulations state that the owner/operator is responsible for airworthiness. But airworthiness is only one element of the safety equation. We must reinforce a culture that everyone is responsible for flight safety. No one is immune. Ethical responsible behavior must be demanded of all. Somehow I perceive that we have lost much of this in recent years.

I notice a minimization and even the omission of ethics in engineering and mechanic curriculums. Why?

Indifference

Aviation has always demanded passion and commitment. If you do not have it, get out! Indifference is dangerous to the aeronautical arts. It is another sure recipe for disaster.

Many contemporary airmen, for some reason, suffer this malaise. Indifference manifests itself in many ways… "it's just a lousy job," "it's not my job," or "no contract no work." Regardless of what we think of our organization, employer, or job at the moment those feelings must be subordinated to the higher responsibility of safety.

Distrust

"In God we trust. Everyone else we check." Indifference has led us to forget this hackneyed aphorism. But safety demands a culture of distrust. We must respect each other but we must also distrust each other. Not in a derogatory sense. Rather as an admission that each of us is capable of error.

Mechanics and engineers who are exposed to airplanes in their daily work must constantly look for what is wrong with the design, what is wrong or out of place on the in-service airplane, what is wrong with maintenance data, procedures, processes. What is broken, leaking, corroded, deformed, chaffing, cracked, etc.? Always assume that something is wrong that will compromise safety.

There you have some of my thoughts. Mostly remember that safety is an individual responsibility, not the responsibility of institutions. Sweep out corners. It's not flashy but it will help to avert disaster. Shakespeare knew it when he talked about horses and nails. Can we afford to forget the lesson?

2

The System

1

Regulatory Basis of Aviation

The United States legislative process

All legislative powers are vested in the Congress of the United States, as granted in Article I, Section I, of the Constitution. The Congress consists of two houses: the Senate and the House of Representatives. Congress meets at least once a year.

Legislative proposals are unbound and diverse. Primary among them are:

- Those instituted by a member of Congress.
- Individuals or citizen groups petitioning members of Congress.
- State legislatures approaching Congress to enact federal laws by passing resolutions that are transmitted to the House and Senate as "memorials."
- Proposals from the executive branch. The Constitution imposes an obligation on the President to report to Congress from time to time on the "state of the union" and to recommend for consideration such measures, as the President considers necessary and expedient. Many of these executive communications follow on the President's message to Congress on the state of the union.
- Executive proposal may also originate in the form of:
    - A message or letter from a member of the President's cabinet (the Secretary of Transportation, for example).
    - A recommendation from the head of an independent agency (the chairman of the National Transportation Safety Board as an example).

- The President transmitting a draft of a proposed bill to the Speaker of the House of Representatives and the president of the Senate.

A sponsoring member introduces proposed legislation to the Congress. It may be entered in either of the two houses and takes the form of a bill or resolution. Bills account for the majority of the legislative proposals before Congress.

A bill is a legislative proposal of a general nature. It proposes either a public or private[1] matter, public bills being the most numerous. The bill is debated, researched, and amended until its wording has been agreed to, in identical form, by both houses of the Congress. It is then submitted to the President for his approval.

A bill becomes the law of the land only after:

- Presidential approval
- Failure by the President to return it with objections to the House in which it originated within 10 days while Congress is in session
- The overriding of a Presidential veto by a two-thirds vote in each house

It does not become law without the President's signature.

Federal law

Legislation enacted by the Congress and signed by the President is first published as a "slip law" (Figure 1-1). This is a printing,[2] in pamphlet form, of an act of Congress and signed by the President (a law).

The slip law is the official publication of the law and is admissible as evidence of the law in all courts of law.[3]

Slip laws, at the end of each session of Congress, are in turn compiled in chronological order into a series of documents known as the *United States Statutes at Large*. This document contains all acts that were passed by Congress. It is the "law of the land."

[1]Private bills are designed to affect or benefit specific individuals or groups of individuals.

[2]The Office of the *Federal Register* publishes slip laws.

[3]See 1 USC 113.

FEDERAL AVIATION ACT OF 1958[1]

Public Law 85-726; 72 Stat. 737

49 U. S. C. App. 1301 et seq.

AN ACT To continue the Civil Aeronautics Board as an agency of the United States to create a Federal Aviation Agency, to provide for the regulation and promotion of civil aviation in such a manner as to best foster its development and safety, and to provide for the safe and efficient use of the airspace by both civil and military aircraft, and for other purposes.

Be it enacted by the Senate and House of Representatives of the United States of America in Congress assembled. That this Act, divided into titles and sections according to the following table of contents, may be cited as the "Federal Aviation Act of 1958":

TABLE OF CONTENTS

Title I – General Provisions

Title II - Civil Aeronautics Board: General Powers of Board[2]

Title III - Organization of Agency AND Powers AND Duties of Administrator

[1] For terminations and transfers of functions of the Civil Aeronautics Board made by the Airline Deregulation Act of 1978 and the Civil Aeronautics Board Sunset Act of 1984, see section 1601.
[2] Title II (other than section 204) ceased to be in effect after January 1, 1985. Section 1601(a)(3).

Fig. 1-1 *Federal slip law.*

Significant laws[4] addressing aviation and identified by their popular names include:

- The Air Commerce Act of 1926 established the first technical regulation of aviation under the Secretary of Commerce. The

[4]Aviation law is federally based, with the states prohibited from regulating rates, routes, or services of any air carrier authorized under the Federal Aviation Act to provide *interstate* air transportation. However, states are not prohibited from enacting consistent laws, or regulating intrastate air carrier activity associated with rates, routes, and services.

act established federal regulations regarding aircraft, airmen, navigational facilities, and the establishment of air traffic regulations. Aircraft were required to be inspected for airworthiness, and were required to have markings placed on the outside of the aircraft for identification. Airmen tests for aeronautical knowledge and physical standards were established. The federal government was required to build new airports, institute regulations that would address aircraft altitude separation, develop and maintain airways and navigational aids. The Department of Commerce Aeronautical Division would be responsible for overseeing and implementing this act. The regulations would be known as the Civil Air Regulations (CARs).

- The Purification Act of 1934 broke up what was essentially a cartel of manufacturers and airlines. During this time, many of the manufacturers of aircraft were also providing passenger service such as Boeing Air Transport. Airlines were separated from manufacturers. This law dissociated manufacturers from passenger travel. It reintroduced competitive bidding as a means for commercial airlines to obtain mail contracts. United Airlines was created from the Boeing Air Transport system to provide passenger travel as we know it today. This act also:
  - Put the U.S. Post Office out of the aviation business and limited the Postal Service to the award of postal contracts but not rate-setting authority.
  - Authorized the Interstate Commerce Commission to set airmail rates.
  - Mandated the Bureau of Air Commerce to the regulation of airways and air safety.
- The Civil Aeronautics Act of 1938 established the Civil Aeronautics Authority, which was separate from the Department of Commerce. In 1940 President Franklin Roosevelt divided the Civil Aeronautics Authority into the Civil Aeronautics Board (CAB) and the Civil Aeronautics Administration (CAA). The CAB was established as an independent agency. It was responsible for issuing and overseeing aircraft and pilot certification and suspension. Additionally, the board issued air carrier route certificates and regulated airline fares. It was also responsible for investigating aircraft accidents.

The CAA was reassigned to the Department of Commerce. It became responsible for aviation technical regulations, airways, navigational facilities, and compliance to aviation regulations. This included air traffic control, airman and aircraft certification, safety enforcement, and airway development. In 1946 Congress added a federal aid airport program to the responsibilities of the Civil Aeronautics Administration.

- The Federal Aviation Act of 1958 redefined technical regulation. It established a new agency known as the Federal Aviation Agency replacing the old Civil Aeronautics Administration. This act granted the FAA exclusive responsibility for the nation's civil-military system of air navigation and air traffic control. Additionally, the former responsibilities of the Civil Aeronautics Administration and the safety rulemaking functions of the Civil Aeronautics Board were transferred to the new agency. The Civil Aeronautics Board retained the responsibility for economic regulation of air carriers and the investigation of aircraft accidents.
- The Independent Safety Board Act of 1974 established the National Transportation Safety Board as an independent agency.
- Air Transportation Security Act of 1974 established security programs and requirements on air carriers and airport operators.
- The Department of Transportation Act of 1976 replaced the FAA with a new organization under the Department of Transportation known as the Federal Aviation Administration.
- The Airline Deregulation Act of 1978 redefined economic regulation. This act phased out the Civil Aeronautics Board (CAB), introduced fare and route competition, and permitted unrestricted entry into the air passenger marketplace by new domestic carriers. Economic regulatory functions not eliminated by the legislation were transferred to the Department of Transportation.
- The Civil Aeronautics Board Act of 1984 led to the demise of full economic regulation, and elimination of the CAB.

The United States Code

Every 6 years the *United States Statutes at Large* are codified in the United States Code (USC) (Figure 1-2). This is a topical consolidation, by subject matter, of the statutes currently in force.

UNITED STATES CODE

1982 EDITION

CONTAINING THE GENERAL AND PERMANENT LAWS
OF THE UNITED STATES, IN FORCE
ON JANUARY 14, 1983

Prepared and published under authority of Title 2, U.S. Code, Section 285b
by the Office of the Law Revision Counsel of the House of Representatives

VOLUME EIGHTEEN

TITLE 46–SHIPPING
TO
TITLE 49–TRANSPORTATION

UNITED STATES
GOVERNMENT PRINTING OFFICE
WASHINGTON : 1983

Fig. 1-2 *United States Code.*

The United States Code does not contain all statutes; for example, private laws or temporary provisions such as budget allocations. For our purposes in this discussion, we may consider the code the law.

The code is divided into 50 titles (Table 1-1) by subject matter. Each title, in turn, is divided into sections. Sections (Figure 1-3) within a title may be further grouped together as subtitles, chapters, sub-chapters, parts, subparts, or divisions. Titles may also have appendices, which may be divided into sections, rules, and/or forms. For our studies, the principal document of the USC is Title 49 Transportation, which contains the majority of the law affecting air carrier maintenance operations. When citing a specific element of the law, I shall refer to the United States Code throughout the remainder of the text as (title number) USC (individual paragraph). Example: 49 USC 44713 is Title 49 of the United States Code, paragraph 44713.

The regulatory process

Within the executive branch of the government, various executive and independent agencies exist to administer the law. Federal laws contain, within their text, the basis for administering the law through one of these executive branch organizations by the adoption of appropriate regulations, policies, procedures, and practices. The dominant organization for our studies is the Federal Aviation Administration (FAA).

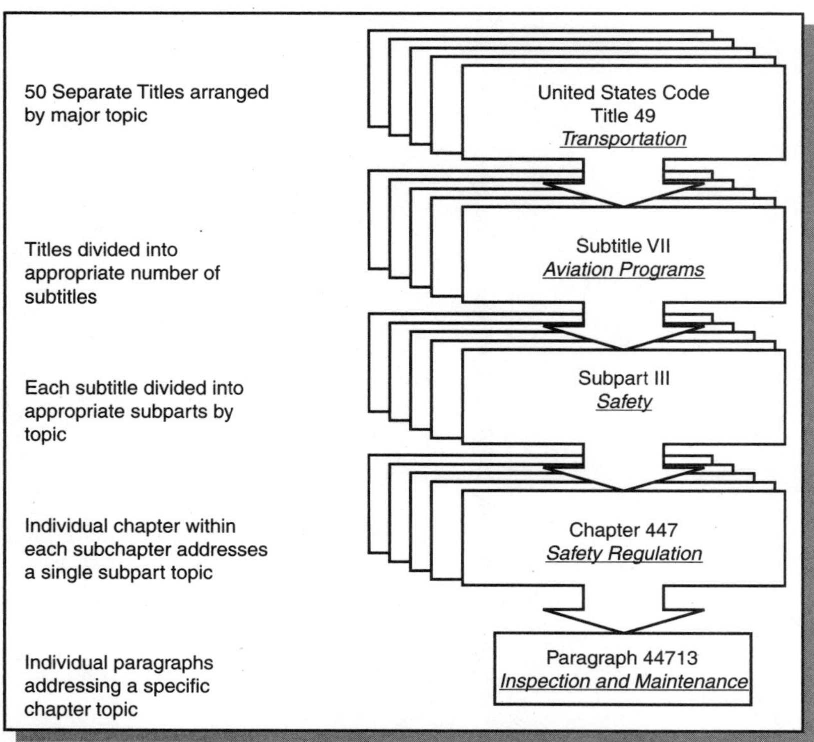

Fig. 1-3 *United States Code organization.*

Table 1-1
United States Code Titles

Title 1 General Provisions	Title 2 The Congress	Title 3 The President	Title 4 Flag and Seal, Seat of Government and the States	Title 5 Government Organization and Employees
Title 6 Unused	Title 7 Agriculture	Title 8 Aliens and Nationality	Title 9 Arbitration	Title 10 Armed Forces
Title 11 Bankruptcy	Title 12 Banks and Banking	Title 13 Census	Title 14 Coast Guard	Title 15 Commerce and Trade
Title 16 Conservation	Title 17 Copyrights	Title 18 Crimes and Criminal Procedure	Title 19 Customs Duties	Title 20 Education
Title 21 Food and Drugs	Title 22 Foreign Relations and Intercourse	Title 23 Highways	Title 24 Hospitals and Asylums	Title 25 Indians

Title 26 Internal Revenue Code	Title 27 Intoxicating Liquors	Title 28 Judiciary and Judicial Procedure	Title 29 Labor	Title 30 Mineral Lands and Mining
Title 31 Money and Finance	Title 32 National Guard	Title 33 Navigation and Navigable Waters	Title 34 Unused	Title 35 Patents
Title 36 Patriotic Societies and Observances	Title 37 Pay and Allowances of the Uniformed Services	Title 38 Veterans Benefits	Title 39 Postal Service	Title 40 Public Buildings, Property and Works
Title 41 Public Contracts	Title 42 The Public Health and Welfare	Title 43 Public Lands	Title 44 Public Printing and Documents	Title 45 Railroads
Title 46 Shipping	Title 47 Telegraphs, Telephones and Radiotelegraphs	Title 48 Territories and Insular Possessions	**Title 49 Transportation**	Title 50 War and National Defense; and Appendix

The establishment of a rule[5] arises from:

- Congressional legislation. Examples are recommendations arising from the 1997 Gore Commission.
- Aviation Rulemaking Advisory Committee (ARAC) activity.
- Technology advances such as the introduction of fly-by-wire or fly-by-light flight control systems.
- Petitions for rulemaking action. Any interested person has the right to petition for the issuance, amendment, or repeal of a rule. These can be as simple as an individual or business petitioning a regulatory agency for an exemption to an existing rule or as complicated as a petition requesting the issuance of new complicated design standards for devices.
- Special or unique operating conditions which arise. A perfect example is the operations of twin engine airplanes in extended overwater flights.
- Lessons learned from accidents or incidents. Petitions for rule changes arising from National Transportation Safety Board recommendations are excellent examples.
- Deficiencies found in certificated products that require correction. This includes the discovery of a potentially hazardous condition unforeseen during initial certification of a product. This is the principal cause of what are called Airworthiness Directives (AD). These are discussed in a later chapter.

The process[6] (Figure 1-4) consists of a draft proposal of a new or amended regulation that is issued by the responsible federal agency known as a Notice of Proposed Rule Making (NPRM) (Figure 1-5).

It is common practice for the regulating agency to involve the public during the drafting of the proposal. Working directly with interested parties does this. One form of this activity is the Aviation Rulemaking Advisory Committee used by the FAA. The draft may

[5]Federal administrative law governs the activity of agencies themselves. See 5 USC 553, Rulemaking.

[6]Administrative law encompasses laws and legal principles governing the administration and regulation of federal government agencies. These agencies are delegated power by Congress to act as agents for the executive. The law is defined in USC Title 5. See 14 CFR Part 11, General Rulemaking Procedures, for detailed rules used by the FAA.

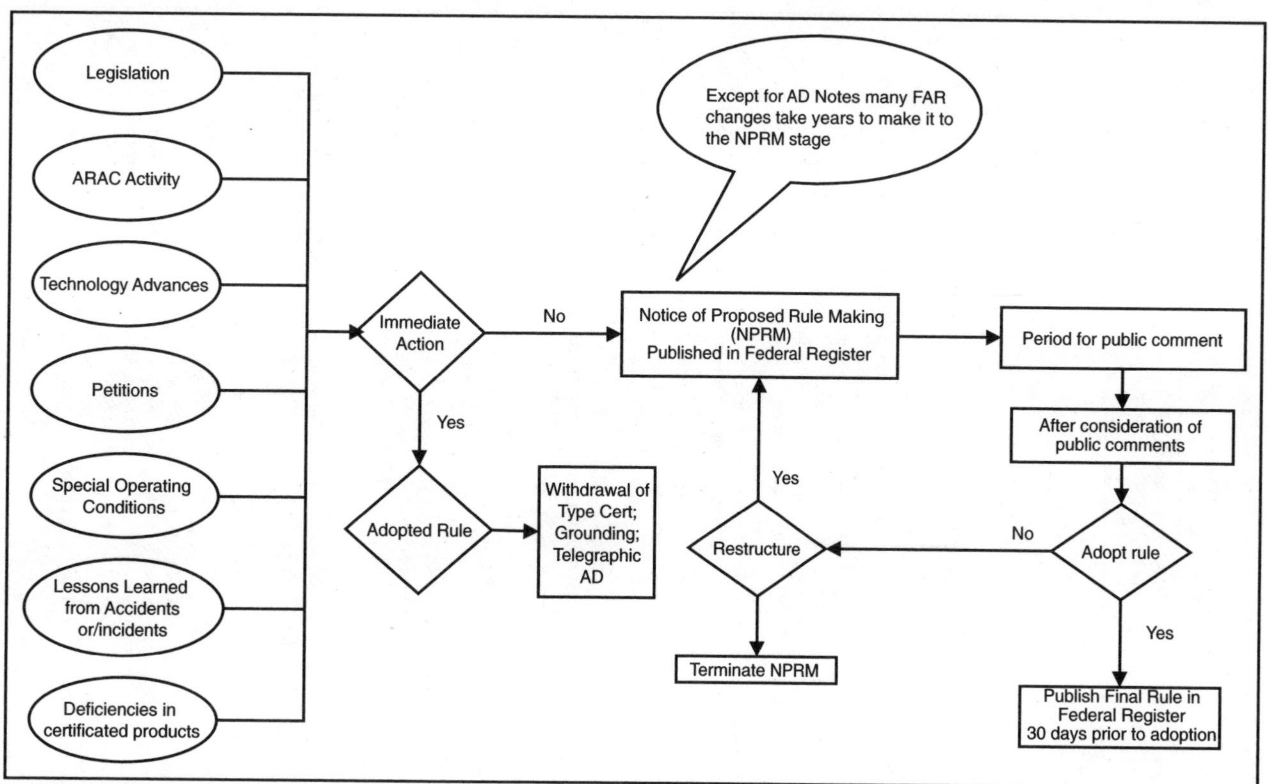

Fig. 1-4 *Regulatory process.*

[Federal Register. May 4, 2000 (Volume 65, Number 87)]
[Proposed Rules]
[Page 25892-25893]
From the Federal Register Online via GPO Access (wais.access.gpo.gov)
[DOCID:fr04my00-14]

Proposed Rules Federal Register

This section of the FEDERAL REGISTER contains notices to the public of the proposed issuance of rules and regulations. The purpose of these notices is to give interested persons an opportunity to participate in the rulemaking prior to the adoption of the final rules.

[(Page 25892)]
DEPARTMENT OF TRANSPORTATION
Federal Aviation Administration
14 CFR Part 39
(Docket No. 99-NE-48-AD]
RIN 212Q-AA64
Airworthiness Directives: General Electric Aircraft Engines CT7 Series Turboprop Engines
AGENCY: Federal Aviation Administration, DOT
ACTION: Notice of proposed rulemaking (NPRM).

SUMMARY: This document proposes the adoption of a new airworthiness directive (AD) that is applicable to certain General Electric Aircraft Engines (GEAE) CT7 series turboprop engines. This proposal would require initial and repetitive inspections of the propeller gearbox (PBG) oil filter impending bypass button (IBB) for extension (popping). This proposal would also require follow-on inspections, maintenance, and replacement actions if the PGB oil filter IBB is popped; and if necessary, replacement of the PGB with a serviceable PGB. In addition, this proposal would require a one-time removal of possibly improperly hardened PGB input pinions and replacement with PGB input pinions that were manufactured using the proper hardening process as terminating action to the repetitive inspections. This proposal is prompted by reports of improperly hardened PGB input pinions. The actions specified by the proposed AD are intended to prevent PGB input pinion failure, which could result in PGB failure and an in-flight engine shutdown.

DATES: Comments must be received by July 3, 2000. ADDRESSES: Submit comments to the Federal Aviation Administration (FAA), New England Region, Office of the Regional Counsel, Attention: Rules Docket No. 99-NE-48-AD, 12 New England Executive Park, Burlington, MA 01803-5299. Comments may also be sent via the Internet using the following address: "9-ane-adcomment@faa.gov." Comments sent via the Internet must contain the docket number in the subject line. Comments may be inspected at this location between 8:00 a.m. and 4:30 p.m., Monday through Friday, except Federal holidays.

The service information referenced in the proposed rule may be obtained from General Electric Aircraft Engines, 1000 Western Ave., Lynn, MA 01910; telephone (781) 594-3140, fax (781) 594-4805. This information may be examined at the FAA, New England Region, Office of the Regional Counsel, 12 New England Executive Park, Burlington, MA.

FOR FURTHER INFORMATION CONTACT: Barbara Caufield, Aerospace Engineer, Engine Certification Office, FAA, Engine and Propeller Directorate, 12 New England Executive Park, Burlington, MA 01803-5299; telephone (781) 238-7173, fax (781) 238-7199.

SUPPLEMENTARY INFORMATION:

Comments invited

Interested persons are invited to participate in the making of the proposed rule by submitting such written data, views, or arguments as they may desire. Communications should identify the Rules Docket number and be submitted to the address specified above. All communications received on or before the closing date for comments, specified above, will be considered before taking action on the proposed rule. The proposals contained in this notice may be changed in light of the comments received.

Comments are specifically invited on the overall regulatory, economic, environmental, and energy aspects of the proposed rule. All comments submitted will be available, both before and after the closing data for comments, in the Rules Docket for examination by interested persons. A report summarizing each FAA–public contact concerned with the substance of this proposal will be filed in the Rules Docket. Commenters wishing the FAA to acknowledge receipt of their comments submitted in response to this notice must submit a self-addressed, stamped postcard on which the following statement is made: "Comments to Docket Number 99-NE-48-AD." The postcard will be date stamped and returned to the commenter. Availability of NPRMs. Any person may obtain a copy of this NPRM by submitting a request to the FAA, New England Region, Office of the Regional Counsel, Rules Docket No. 99-NE-48-AD, 12 New England Executive Park, Burlington, MA 01803-5299. Discussion. The Federal Aviation Administration (FAA) has received reports of improperly hardened propeller gearbox (PGB) input pinions installed on General Electric Aircraft Engines (GEAE) CT7 series turboprop engines.

Fig. 1-5 *Typical NPRM.*

take several years of debate and processing before its release as an NPRM.[7]

The document is issued a docket number[8] and the contact name of an individual within the agency along with an address and phone number.

The notice contains a standard format including:

- A statement of the time, place, and nature of public rule-making proceedings.
- Reference to the legal authority under which the rule is proposed.
- The terms or substance of the proposed rule or a description of the subjects and issues involved. This is not applicable
  - To interpretative rules, general statements of policy, or rules of agency organization, procedure, or practice.
  - When the agency for good cause finds that notice and public procedure thereon is impracticable, unnecessary, or contrary to the public interest.

After an NPRM is promulgated, interested persons are given an opportunity to participate in the rulemaking through the submission of written data, views, or arguments with or without opportunity for oral presentation. Any person or organization may comment on an NPRM. For each notice, detailed instructions on how, when, and where a viewpoint may be expressed are provided. Many agencies accept comments via e-mail. Comment periods vary, but are usually 30, 60, or 90 days.

When the final rule is adopted, a statement by the Administrator must address the significant issues raised in comments, reasons for their disposition, and a discussion of any changes made in response to them. This statement includes the Administrator's review of the public comment including clarifying comments regarding the mean-

[7]The FAA is a huge bureaucracy. Unfortunately the time to adopt a regulation or change it seems to take longer and longer. This is a common complaint of both regulators and the regulated. Furthermore once a rule is adopted, change is slow in coming (it seems impossible) even if the rule turns out to be defective. This leads to the frequent use of other documents, not subject to the rulemaking process, such as policy statements, memoranda, Advisory Circulars, etc., taking on the role of de facto regulations.

[8]Docket numbers identify documents within the agency's internal filing system. The docket contains all documents, testimony, comments, exhibits, etc., relating to the NPRM. It is available for public inspection.

ing, background, and intent of the rule being adopted. This is an important source of historical information when researching the meaning and intent of a rule.

The required publication or service of a rule must be made not less than 30 days before its effective date, with the following exceptions:

- A rule that grants or recognizes an exemption or relieves a restriction.
- Interpretative rules and statements of policy.
- The reason(s) are published with the rule except administrative changes or editorial changes to correct typographical errors in an existing rule.

Frequently the final rule carries a compliance date or effectivity should it require some direct action on the part of those agents/agencies affected. An excellent example is Airworthiness Directives.[9] The condition, giving rise to the adoption of an immediate adopted AD rule, may be so serious that it requires immediate action without a period for comment.

Regulatory evaluations

Changes to federal regulations must undergo several evaluations. I shall not burden you with the details. For examples of how convoluted federal regulations can become consider the following:

1. Executive Order 12866 directs that each federal agency shall propose or adopt a regulation only upon a reasoned determination that the benefits of the intended regulation justify its costs.
2. The Regulatory Flexibility Act of 1980 requires agencies to analyze the economic impact of regulatory changes on small entities.
3. The Office of Management and Budget (OMB) directs agencies to assess the effect of regulatory changes on international trade.
4. The Un-funded Mandates Reform Act of 1995 (Pub. L. 104-4) requires agencies to assess the costs, benefits, and other effects of proposed or final rules that include a federal mandate

[9]Airworthiness Directives address items directly affecting flight safety. We shall discuss them in a greater detail in a following chapter.

likely to cause state, local, or tribal governments to expend funds as a result.

5. An International Trade Impact Assessment affects the impact of trade for U.S. firms doing business in foreign countries and foreign firms doing business in the United States.
6. Executive Order 13132, Federalism. The FAA must determine if the rule will have a substantial direct effect on the states, on the relationship between the national government and the states, or on the distribution of power and responsibilities among the various levels of government.
7. Regulations Affecting Intrastate Aviation in Alaska Section 1205 of the FAA Reauthorization Act of 1996.
8. The energy impact of the rule must be assessed in accordance with the Energy Policy and Conservation Act (EPCA) Pub. L. 94-163.

And you think airline operation and its regulation are simple!

Aviation Rulemaking Advisory Committee (ARAC)[10]

In the late 1980s, the Secretary of Transportation and the Administrator created a task force on FAA reform to recommend improvements that could be made in the operations within the administration.

One result of that group's activity was the establishment of an advisory committee to serve as a forum for the FAA to gain input from outside the government on major regulatory issues. The intent is:

- Take advantage of industry technical expertise and experience
- Resolve any issues or controversies in an open forum before beginning formal rulemaking
- Broaden public participation in the regulatory process
- Minimize congressional inclination to mandate rulemaking

The committee provides advice and recommendations to the Administrator with respect to proposed and existing rules that should

[10]The existence of ARAC arises from The Federal Advisory Committee Act of 1972 (Pub. L. 92-463). See Title 5, USC Appendix II.

be revised or eliminated. Activities of the committee may not circumvent public rulemaking. Any FAA/committee activity on the development of proposed rules is disclosed in the public docket. Committee meetings are open to the public, including participation in the activities. Neither the committee nor any of its subgroups may assign a task without prior approval by the FAA. The committee's present charter expired in February 2000.

The committee consists of 66 member organizations selected by the FAA (Figure 1-6). They are intended to be representative of the aviation community and appropriate public interest groups. They serve without compensation. Issues presently being addressed by the ARAC are summarized in Figure 1-7.

The *Federal Register*

The *Federal Register* (Figure 1-8), published every business day, is the official venue for rules, proposed rules, and notices of federal organizations, including executive orders and Presidential documents. It is a legal newspaper published by the National Archives and Records Administration. It is distributed in paper, on microfiche, and also on the World Wide Web.

Each issue of the *Federal Register* is organized into four major categories:

- *Rules and Regulations.* This category contains final rules and regulations—regulatory documents having general applicability and legal effect. Most rules are keyed to and codified in the Code of Federal Regulations (CFR). A document that amends text includes the changes to the CFR and states the effective date for the change. This section also contains interim rules that are issued without prior notice[11] and are effective immediately. An interim rule is designed to respond to an emergency situation and is usually followed by a final rule. This section includes documents that have no regulatory text and also do not amend the CFR. These either affect an agency's administrative handling of its regulations or are of continuing interest to the public in dealing with an agency. In this classification are general policy statements and interpretations of agency regulations.

[11]These are commonly known as adopted rules.

1. AECMA
2. Aeronautical Repair Station Association
3. Aerospace Industries Association of America, Inc.
4. Aerospace Industries Association of Canada
5. Aerospace Medical Association
6. Air Line Pilots Association
7. Air Traffic Control Association
8. Air Transport Association of America
9. Airbus Industrie
10. Aircraft Electronics Association
11. Aircraft Owners and Pilots Association
12. Airline Dispatchers Federation
13. Airline Suppliers Association
14. Airport Consultants Council
15. Airports Council International, N.A.
16. Alaska Air Carriers Association
17. Allied Pilots Association
18. American Association of Airport Executives
19. American Helicopter Society
20. Association of Air Medical Services
21. Association of European Airlines
22. Association of Flight Attendants
23. Association of Professional Flight Attendants
24. Aviation Consumer Action Project
25. Aviation Insurance Association
26. Aviation Technican Education Council
27. Balloon Federation of America
28. Boeing Commercial Airplane Group
29. Cargo Airline Association
30. Equipment Leasing Association of America
31. Experimental Aircraft Association
32. Flight Dispatchers, Meteorologists, & Operations Specialists Union
33. Flight Safety Foundation
34. General Aviation Manufacturers Association
35. Helicopter Association International
36. Independent Pilots Association
37. International Airline Passengers Association
38. International Association of Machinists and Aerospace Workers
39. International Foundation for Airline Passengers
40. International Society of Aviation Maintenance Professionals
41. Joint Aviation Authorities Secretariat
42. Light Aircraft Manufacturers Association
43. National Aeronautic Association
44. National Agricultural Aviation Association
45. National Air Carrier Association, Inc.
46. National Air Transportation Association, Inc.
47. National Aircraft Finance Association
48. National Association of Flight Instructors
49. National Association of Trade and Technical Schools
50. National Association of State Aviation Officials
51. National Business Aviation Association, Inc.
52. National Fire Protection Association
53. National Organization to Insure a Sound Controlled Environment
54. Parachute Industry Association
55. Popular Rotorcraft Association, Inc.
56. Professional Aviation Maintenance Association
57. Public Citizen
58. Regional Airline Association
59. Small Aircraft Manufacturers Association
60. Teamsters Airline Division
61. The Soaring Society of America, Inc.
62. Transport Canada
63. United States Parachute Association
64. United States Ultralight Association, Inc.
65. University Aviation Association
66. Used Aircraft Certification Conformity Committee

Fig. 1-6 *ARAC membership.*

— Air Carrier Operations Issues
— Training and Qualifications Issues
— Air Carrier/General Aviation Maintenance Issues
— General Aviation and Business Airplane Issues
— Air Traffic Issues
— Noise Certification Issues
— Emergency Evacuation Issues
— Aircraft Certification Procedures Issues
— General Aviation Operations Issues
— Airport Certification Issues
— Rotorcraft Issues
— Transport Airplane and Engine Issues
— Fuel Tank Harmonization Working Group (arise from TWA 800)
— Fractional Ownership

Fig. 1-7 *ARAC issues.*

- *Proposed Rules* (NPRM). This section contains public notice of contemplated rules and regulations. This gives interested persons an opportunity to participate in the rulemaking process. Proposals suggest changes to agency regulations and solicit public comment. Agencies further use the notice and comment process to stay in contact with constituents, soliciting views on procedural rules, interpretative rules and agency policy issues, and program issues; for instance the necessity for and adequacy of anticipated regulatory action. Finally, documents relating to previously published proposals are included in this section. These address extension of a comment period, notice of a public hearing, addition of supplemental information, withdrawal of a proposed rule, or correction of a previously published proposal.
- *Notices.* These include announcements of hearings and investigations, committee meetings, agency decisions and rulings, delegations of authority, issuance or revocation of licenses, grant application deadlines, availability of environmental impact statements, filing of petitions[12] and applications, and agency statements of organization and functions.
- *Presidential Documents.* This section contains papers signed by the President. Presidential documents include

[12] 14 CFR Part 11.25(a) "Any interested person may petition the Administrator to issue, amend, or repeal a rule...."

Friday,
September 15, 2000

Federal Register

Part IV

Department of Transportation

Federal Aviation Administration

14 CFR Part 121 et al.
Service Difficulty Reports; Final Rule

Fig. 1-8 *Federal Register.*

proclamations and executive orders as well as other materials such as determinations, letters, memoranda, and reorganization plans.

Code of the Federal Regulations (CFR)

The purpose of the CFR (Figure 1-9) is to present the official, complete text of agency regulations in one organized publication. The code is an annual codification of the general and permanent rules[13] adopted by various executive departments and agencies of the federal government.[14] It is organized into 50 titles that represent broad subject areas (Table 1-2).

Each title is divided (Figure 1-10) into chapters. Chapters in turn are further subdivided into subchapters and then parts covering specific regulatory areas. Large chapters may be further subdivided into subparts. When citing a specific regulation, I shall refer to the Code of the Federal Regulations throughout the remainder of the text as (title number) CFR (individual part.paragraph). Example: Federal Aviation Regulation (FAR) Part 121.703 is cited as 14 CFR Part 121.703.

The CFR is published jointly by the Office of the Federal Register and the Government Printing Office (GPO). Paper or microfiche versions are available through subscription. It is also available on the World Wide Web. Furthermore, commercially prepared customized paper and electronic versions of the document are available to meet the needs of specific users.

It is updated and published once a year in print, fiche, and on-line formats. The approximately 200 CFR volumes are revised at least once a year on a quarterly basis (Table 1-3).

Federal aviation regulations

For our studies, the principal document of the CFR is Title 14 Aeronautics and Space. Chapter 1, Parts 1 through 199,[15] is issued by the Federal Aviation Administration. It is the principal body of regulation that affects maintenance and flight operations. It contains the technical rules governing civil aviation, popularly called the FAR. Approximately 106 of these parts are active regulations. The remainder are unassigned. The dominant air carrier regulations are 14 CFR Part 119, Certification: Air Carriers and Commercial Operators, and 14

[13] Originally published in the *Federal Register*.

[14] The CFR is keyed to and kept up-to-date by the *Federal Register*. These two publications must be used together to determine the latest version of a given rule. For example, Title 14 is updated each January 1. Thus at a later time in the same year it would be necessary to review the *Federal Register* issues from the first of the year for any changes occurring to a rule since the last revision of Title 14.

[15] Chapter 2, Parts 200 through 399, relates to the economic regulations that remain after deregulation. The office of the secretary of transportation issues these. They establish aviation proceedings relating to economic as well as administrative regulation. An example is regulations for the issuance of certificates of public convenience and necessity.

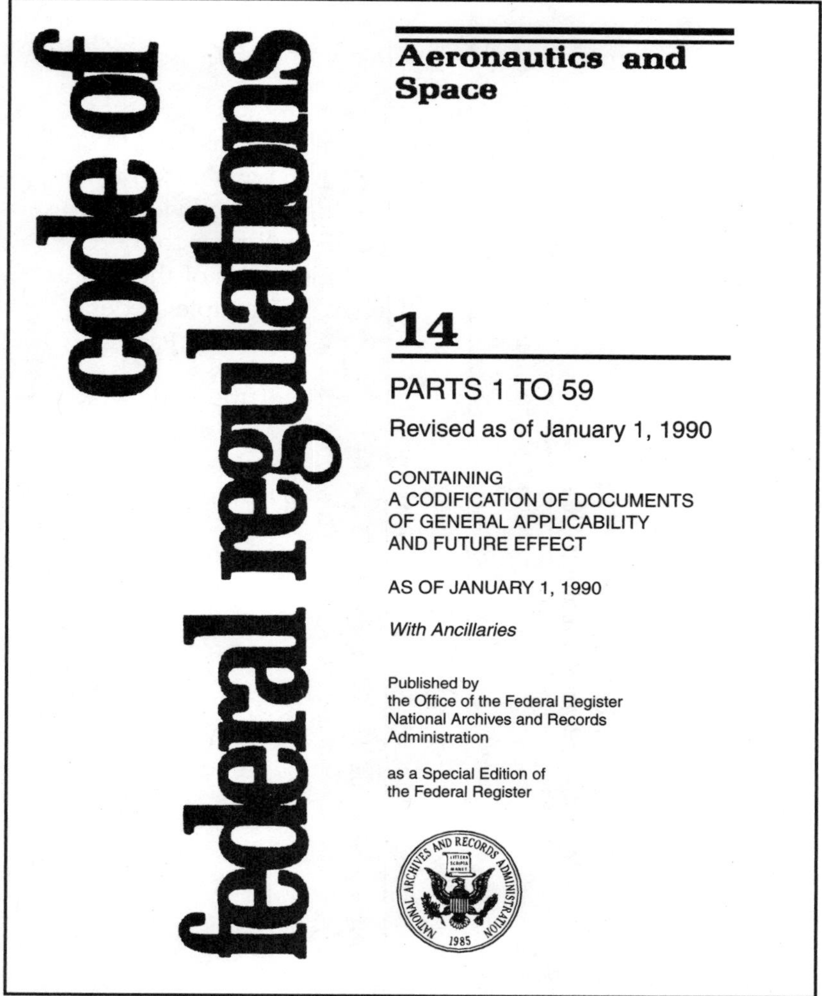

Fig. 1-9 *Code of the Federal Regulations.*

CFR Part 121, Operating Requirements: Domestic, Flag, and Supplemental Operations.

This is not the only body of regulation affecting maintenance operations. As shall be seen in subsequent chapters, numerous other titles that contain regulations affecting the way an operator does business are contained throughout the CFR, for example, some airplanes have depleted-uranium rudder balance weights. The

Table 1-2
Code of the Federal Regulations Titles

Title 1 General Provisions	Title 2 [Reserved]	Title 3 The President	Title 4 Accounts	Title 5 Administrative Personnel
Title 6 [Reserved]	Title 7 Agriculture	Title 8 Aliens and Nationality	Title 9 Animals and Animal Products	Title 10 Energy
Title 11 Federal Elections	Title 12 Banks and Banking	Title 13 Business Credit and Assistance	**Title 14 Aeronautics and Space**	Title 15 Commerce and Trade
Title 16 Commercial Practices	Title 17 Commodity and Securities Exchanges	Title 18 Conservation of Water and Power Resources	Title 19 Customs Duties	Title 20 Employee's Benefits
Title 21 Food and Drugs	Title 22 Foreign Relations	Title 23 Highways	Title 24 Housing and Urban Development	Title 25 Indians

Title 26 Internal Revenue	Title 27 Alcohol, Tobacco and Firearms	Title 28 Judicial Administration	Title 29 Labor	Title 30 Mineral Resources
Title 31 Money and Finance: Treasury	Title 32 National Defense	Title 33 Navigation and Navigable Waters	Title 34 Education	Title 35 Panama Canal
Title 36 Parks, Forests and Public Property	Title 37 Patents, Trademarks and Copyrights	Title 38 Pensions, Bonuses, and Veterans' Relief	Title 39 Postal Service	Title 40 Protec-tion of Environment
Title 41 Public Contracts and Property Management	Title 42 Public Health	Title 43 Public Lands; Interior	Title 44 Emergency Management and Assistance	Title 45 Public Welfare
Title 46 Shipping	Title 47 Telecommunications	Title 48 Federal Acquisition Regulations System	Title 49 Transportation	Title 50 Wildlife and Fisheries

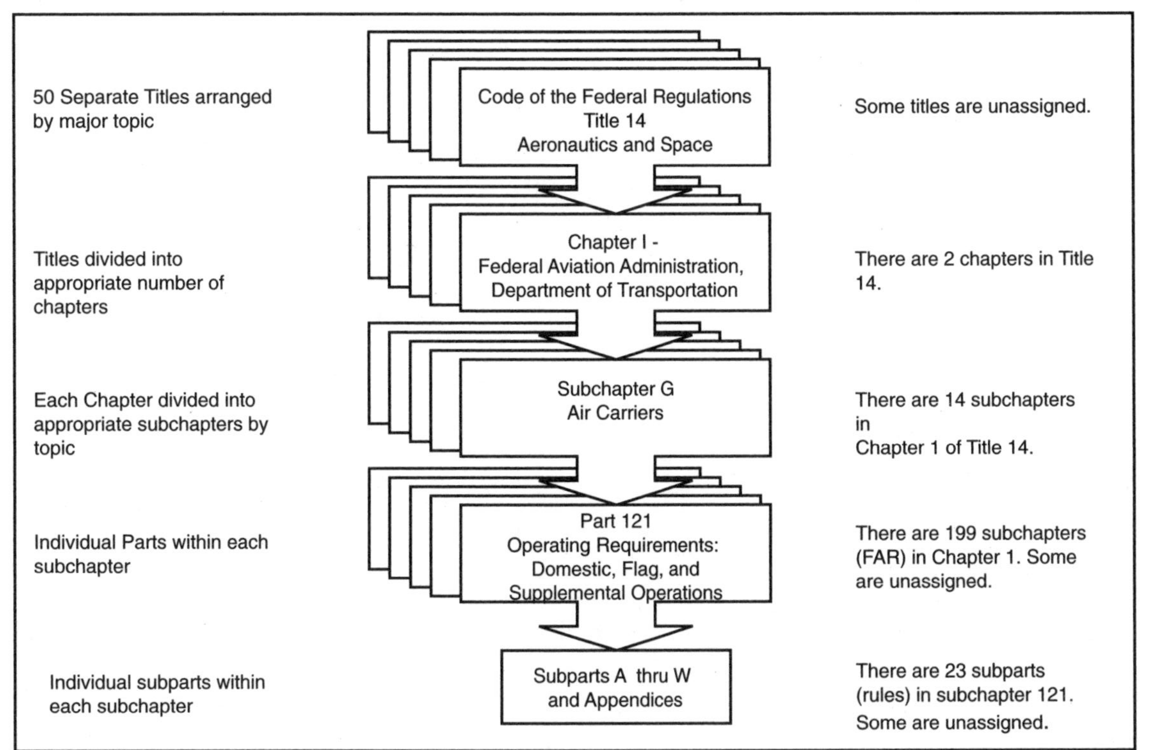

Fig. 1-10 *Code of the Federal Regulations Organization.*

Table 1-3
Code of The Federal Regulations—Revision Cycle

Titles 1–16	As of January 1
Titles 17–27	As of April 1
Titles 28–41	As of July 1
Titles 42–50	As of October 1

handling and disposal of these materials are governed by the Nuclear Regulatory Commission regulations.

Old civil air regulations versus the FAR

The Civil Aviation Regulations (CAR) and Civil Aviation Manuals (CAM) became outmoded with the rapid growth of air transportation and the introduction of turbojet transport category airplanes in the 1950s. Recodification of the Civil Aviation Regulations began in 1961 and was completed in 1964 with the adoption of the Federal Aviation Regulations (FAR). The Civil Aeronautics Manuals were replaced by documents called Advisory Circulars.

Although most of the old CARs have been replaced, some retain applicability. For example, the CAR 4b, Airworthiness—Transport Category Airplanes, applies to airplanes originally certified under that regulation such as the 707, DC-8, 727, and so forth.

Exemption to regulations

Conditions occasionally arise that reveal a given regulation as not suitable to a particular operation or activity. Under these circumstances, it is possible to seek relief from the rule through an exemption. Petitions for exemption are published in the *Federal Register* (Fig. 1-11).

Any person may petition for an exemption.[16] Excellent examples are petitions for relief to time limits imposed in a regulation. Detailed procedures and rules for exemption are contained in 14 CFR Part 11.25.

[16]See 49 USC 40109.

[Federal Register; May 10, 2000 (Volume 66, Number 91)]

[Notices]

[Page 30185-30187]

From the Federal Register Online via GPO Access [wais.access.gpo.gov][DOCID.fr1Omy00-147]

DEPARTMENT OF TRANSPORTATION

Federal Aviation Administration

[Summary Notice No. PE-2000-17]

Petitions for Exemption; Summary of Petitions Received; Dispositions of Petitions Issued

AGENCY: Federal Aviation Administration (FAA), DOT.

ACTION: Notice of petitions for exemption received and of dispositions of prior petitions.

SUMMARY: Pursuant to FAA's rulemaking provisions governing the application, processing, and disposition of petitions for exemption (14 CFR Part 11), this notice contains a summary of certain petitions seeking relief from specified requirements of the Federal Aviation Regulations (14 CFR Chapter 1), dispositions of certain petitions previously received, and corrections. The purpose of this notice is to improve the public's awareness of, and participation in, this aspect of FAA's regulatory activities. Neither publication of this notice nor the inclusion or omission of information in the summary is intended to affect the legal status of any petition or its final disposition.

DATES: Comments on petitions received must identify the petition docket number involved and must be received on or before May 31, 2000.

ADDRESSES: Send comments on any petition in triplicate to: Federal Aviation Administration, Office of the Chief Counsel, Attn: Rule Docket (AGC-200, Petition Docket No. ______, 800 Independence Avenue, S.W., Washington, DC 20591.

Comments may also be sent electronically to the following Internet address: 9-NPRM-ombs@faa.gov.

The petition, any comments received, and a copy of any final disposition are filed in the assigned regulatory docket and are available for examination in the Rules Docket (AGC-200), Room 915G. FAA Headquarters Building (FOB 10A), 800 Independence Avenue, SW., Washington, DC 20591; telephone (202) 267-3132.

FOR FURTHER INFORMATION CONTACT: Cherie Jack (202) 267-7271. Forest Rawls (202) 267-8033, or Vanessa Wilkins (202) 267-8029 Office of Rulemaking (ARM-1), Federal Aviation Administration, 800 Independence Avenue, SW., Washington, DC 20591.

This notice is published pursuant to paragraphs (c), (e), and (g) of Sec. 11.27 of Part 11 of the Federal Aviation Regulations (14 CFR Part 11).

Issued in Washington, DC., on May 5, 2000. Donald P. Byrne, Assistant Chief Counsel for Regulations.

Petitions for Exemption

Docket No. 29900. Petitioner: Atlantic Coast Airlines and Trans State Airlines. Section of the FAR Affected: 14 CFR 121.344(d)(1). Description of Relief Sought. To permit ASA and TSA to operate its Jetstream-41 (J-41) aircraft without meeting the requirements of 14 CFR 121.344(d)(1).

Docket No. 29941. Petitioner: Hawaiian Airlines, Inc. Section of the FAR Affected: 14 CFR 25.857(c), 85.858, 121.314(c). Description of Relief Sought: To allow Hawaiian Airlines to operate, until May 15, 2001, on DC10-10 airspace beyond the cargo compartment modification deadline of March 30, 2001.

Docket No. 29981. Petitioner: Delta Air Lines, Inc. Section of the FAR Affected: 14 CFR 25.857(o), 26.858, 121.314(c). Description of Relief Sought: To permit Delta Air Lines to operate, until September 20, 2001, nine L-1011 airplanes beyond the cargo compartment modification deadline of March 19, 2001.

Dispositions of Petitions Docket No. 28419, Petitioner: United Parcel Service. Section of the FAR Affected: 14 CFR 121.433(c)(1)(iii), 121.440(s), [(Page 30183)]; 121.441(a)(1) and (b)(1), and appendix F to part 121. Description of Relief Sought/Disposition: To permit UPS to combine recurrent flight and ground training and proficiency checks for UPS's pilots in command, seconds in command, and flight engineers in a single annual training and proficiency evaluation program. Grant 04/06/2000.

Exemption No. 64348. Docket No. 29883, Petitioner: Embry-Riddle Aeronautical University. Section of the FAR Affected: 14 CFR 61.65(a)(1). Description of Relief Sought/Disposition. To permit ERAU to permit students enrolled in ERAU's AGATE 111 to take concurrently the private pilot and instrument rating practical test, subject to certain conditions and limitations. Grant, 04/14/2000.

Fig. 1-11 *Typical petition for exemption.*

Working with the rules

It is essential to acquire tools that allow you to intelligently discuss and in some cases argue actions proposed or taken by regulatory agencies. This includes the ability to defend against alleged violation of the regulation.

You must know how to debate and negotiate. When engaged in either activity, the following principles, although well known, bear restating:

1. Know well the law and the rule that governs your activity.
2. You must understand your argument. Also, you should have a detailed knowledge of your opponents' argument that is better than their own knowledge of it. Preparation, in detail, is essential.
3. It is not enough to know the problem. It is paramount to know who owns the problem. This premise avoids much wasted effort.
4. Any debate is conducted to convince an opponent of one's position. It is not to irritate. Avoid inflammatory arguments and positions.
5. Principles before personalities. Base an argument upon facts and data, not emotional appeals.

The regulations don't have to make sense, they just are. Accept the rules as they exist. Unquestionably, bureaucratic fog and lack of substantive need burden many FARs. Many of them are out of date. This gives rise to the oft-expressed statement, "The FAR don't make sense." Unfortunately, this expression is frequently used as an apology for ignorance of the system and its perturbations rather than as a statement of frustration with the rules.

Regulations are not a random body of government obfuscation. One must understand not only the content of the rules, but also how they came about. This frequently affects insight into the application of the rule. The existing body of rules[17] arises from experience gained since the beginning of powered flight. Rules frequently bespeak of many hard lessons, arising, unfortunately, from sometimes tragic experience. It is a body of experience gained from the Post Office's airmail system of the 1920s through

[17]Although occasionally tainted by politics, legal nitpicking, and pedantic posturing.

the latest adventures with digital airplanes. It is a system, which must be understood from the perspective of history as well as technology. What gave rise to a rule's existence? Always look to the story behind the rule. This may require tracing the genesis of the rule using the finding aids of the *Federal Register.*

Understand cultural influences upon the system. Notwithstanding public input and industry participation in the process, the law and the regulation are ultimately the product of the political and legal systems, not the aeronautical arts. Rules are written and ultimately interpreted by lawyers, not engineers, mechanics, or pilots. But armed with an understanding, albeit minimal, of the world of lawyers and politicians you will begin to understand how to work within the system.

Understand well, "Rules are rules." I mean that the rule is not what one individual says it is. Don't allow yourself to be blinded by local interpretations and practices of individual inspectors and offices; advice that should also be applied to one's own perceptions and prejudices regarding a rule. The meaning of the rule is established by:

- The Administrator issuing legal opinions. These follow a rules adoption, frequently the product of enforcement action or of inquiries made to the Administrator. They are, however, opinions and may or may not carry the force of law.[18]
- Administrative Law Judges[19] issuing legal opinions arising from certificate action appeals to the NTSB

Further, it is helpful toward understanding what the rule means to review the Administrator's comments[20] when adopting a rule. These have particular merit in the years after the regulation's adoption. The passage of time frequently results in the practical application of the rule to take on a character that is not in conformity to the original

[18]There are many orders, bulletins, policies, internal memoranda, and other miscellaneous documents that are issued by the FAA. These documents are invalid if they conflict with regulation or law. It is not an uncommon practice for FAA employees to quote them as if they have, in fact, the force of regulation and law. Always determine if they do.

[19]The position of the Administrative Law Judge (ALJ), originally called hearing examiner, was created by the Administrative Procedure Act of 1946, Public Law 79-404. The act ensured fairness and due process in federal agency rulemaking and adjudication proceedings. It provided those parties, whose affairs are controlled or regulated by agencies of the federal government, an opportunity for a formal hearing on the record before an impartial hearing officer.

[20]These are found in the *Federal Register* when the final rule is published.

meaning and intent of the rule. This results, unfortunately, in de facto revision to the rule by well-meaning uniformed individuals. The consequence is unwarranted regulatory action or nonuniform enforcement of the regulation.

The determination of the validity of a given rule with the enabling law is the purview of the federal judiciary, notably The United States Court of Appeals.

Any regulatory system is contentious. This is part of the process, which helps level the playing field. Aviation is continually embroiled in arguments regarding the necessity for or the meaning of a given proposed or existing regulation. At the center of these debates is the recognition that law and regulation are not synonymous. Regulation is enabled by and proceeds from the law. It is only valid to the extent that it is authorized by the law and is within the law's meaning and intent. Regulation, policies, and procedures issued by the Administrator, to be valid, must meet this test: no law, then no regulation. It is not uncommon, then, to see in the trade press that someone has filed suit in federal court against the Administrator. In the main these activities claim that a given regulation, policy, or activity is not authorized by the law, and so on.

Finally, you must constantly scrutinize developments in the regulatory and legislative environments. Learn to review the *Federal Register* regularly.

Enforceability of regulations

Rules must contain mandatory or prohibitive language to be enforceable. About half the FAR are not enforceable. They confer authority or responsibility. They are definitive or explanatory. If there is no regulation against it, a person can do it and not be in violation. "What is not forbidden is permitted."

There are six general categories of wordings in regulations.

- *Mandatory.* "shall..." or "must..."
- *Prohibitive.* "No person may..." or "A person may not..."
- *Conditionally mandatory.* "Each person shall..." or "however..."
- *Conditionally prohibitive.* "No person may, except..." or "unless..."

- *Verbiage conferring authority or responsibility.* "...is responsible..."
- *Definitive or explanatory.* "This part prescribes..."

Enforcement action

"...the Administrator may conduct investigations, hold hearings, issue subpoenas, require the production of relevant documents, records, and property, and take evidence and depositions."[21]

Each record, document, or report that the Federal Aviation Regulations requires may be used in any investigation conducted and may be used in any civil penalty action, certificate action, or other legal proceeding.

Readers are advised to become familiar with 14 CFR Part 13, Enforcement Procedures.

International aviation—The Chicago Conference

Beginning during the mid-1930s, technical, political, economic, and legal issues were recognized as central to the development of international commercial air operations. Economic trade issues including rates and tariffs, levels of service, and flight frequencies were unanswered. Which airlines from what country could carry traffic between points within each other's respective territories?

Sovereignty, operating rights between countries, the recognition of property rights in aircraft, damage done by aircraft to third parties on the surface, the liability of the air carrier to its passengers, crimes committed on board aircraft all required answers.

Technical questions also developed. Who and what defines airworthiness? How is air traffic controlled? How do agencies communicate? Are there common procedures?

The outbreak of World War II delayed this work. However, by late 1944, activity again began. The Chicago Conference addressed these questions with mixed success.

[21] 14 CFR Part 13.3(a).

Air service tariffs

The question of tariff setting and control was at first reserved by individual countries. This was unsuccessful for a number of practical as well as political reasons. The solution developed at the Chicago Conference was to empower the International Air Transport Association (IATA) to fulfill the function.[22] IATA regularly holds regional traffic conferences[23] to set rates and service levels over routes. Activities, which are subject to appropriate government approval, include:

- International tariffs
- Levels of service offered by members
- Interlining agreements or the transfer of revenue from one operator to another
- Settlement of accounts through the IATA clearinghouse
- Proration, i.e., revenue division
- Debt settlement
- Technical issues relating to ticketing and documentation such as airway bills
- Engineering technical issues common to all carriers

Air service agreements—bilaterals

The framework of who flies where and how frequently was established at the Chicago Conference. Countries negotiate bilateral agreements for air transport services. These agreements establish the traffic rights of the airlines offering service in international markets. After the agreement, the countries award their airline(s) the right to offer those services.

Traffic rights determine:

- Which routes airlines can serve between the countries and to third countries.
- What services they can provide (e.g., scheduled or charter).
- How many airlines from each country can fly the routes.
- How frequently they can offer flights.

The first bilateral agreement was negotiated between the United States and the United Kingdom in 1946. This agreement, known as

[22]United States carriers have been granted immunity from antitrust law.

[23]At present 100 members participate in tariff-setting activities.

the Bermuda Agreement, has served as the model for negotiating operating rights since that time.

Responsibility for negotiating bilateral agreements on behalf of the United States rests with the State Department and the Department of Transportation. The Office of Aviation and International Affairs of the Department of Transportation provides aviation policy and expertise to the negotiations. The State Department chairs negotiations with foreign governments and coordinates DOT's international aviation policies/actions with overall U.S. foreign policy.

The United States presently has 72 bilateral aviation agreements covering air services to 107 countries.

The freedoms of the air

As a part of the Chicago Conference, participants accepted five freedoms of the air.[24] The freedoms recognize sovereignty and the economics of international air commerce. They are not really freedoms but rather negotiated rights. The first two define technical operating rights. They are normally granted together. They are frequently granted outside the formal conduct of bilateral negotiations. The procedures used by countries to grant first and second freedoms are variable; for example, Canada grants blanket first freedoms to the United States. Mexico requires individual application for rights for general aviation. Clearly these two rights directly affect both flight and maintenance operations. The remaining freedoms apply to economic issues.

The remaining third, fourth, and fifth freedoms identify economic operating rights, which are the product of bilateral negotiations. Third and fourth freedom rights are usually simultaneously granted during bilateral negotiations.

The first freedom

This is the right of an aircraft of registry in country A to overfly the territory of country B without landing. For example, United Airlines is granted first freedom by Canada. United can overfly Canada.

The second freedom

Second freedom rights grant an aircraft of registry in country A to land in the territory of country B for technical reasons only. For

[24]Lord Beaverbrook originally proposed the freedoms during the Dominion Conference in 1943.

example, Lufthansa is granted second freedom by the Emirates. Lufthansa has the right to stop in Dubai en route to Singapore to refuel, for crew change, and so forth. It may neither discharge nor take aboard revenue passengers or cargo.

The third freedom

Third freedom rights grant an airline from country A to accept revenue in country A for transport to country B. For example, British Airways has the right to sell a passenger ticket in London for transport to Brussels, Belgium.

The fourth freedom

Fourth freedom is the inverse of the third freedom right. It is the authority of an airline of country A to accept revenue payloads in country B for transport to country B. For example, British Airways has the right to sell a ticket in Brussels, Belgium, for London.

The fifth freedom

Fifth freedom rights grant authority for an airline from country A to accept revenue payloads in country B for transport to country C. For example, Air France has the right to accept cargo in Rio de Janeiro, Brazil, for shipment to Dakar, Senegal.

Cabotage

Cabotage[25] is not a part of the five freedoms. Its origins are in maritime law. In economic terms, it is the right of a vessel of registry in country A to accept revenue payload in country B and transport that payload to another port in country B. For example, Air Canada sells a ticket in Cleveland for Los Angeles. Cabotage is in this sense very rarely granted.

Air service agreements—Open Skies

Current United States policy promotes competition and economic deregulation in international aviation markets. It seeks an open market structure throughout the world. Traditional bilateral agreements are being phased out.

"Open Skies" agreements, as opposed to restrictive bilateral agreements, allow unrestricted international air service between

[25]The *Oxford English Dictionary* defines cabotage as the "coast carrying trade by sea." In aeronautical terms, it means the right of a country to operate air traffic within its territory.

participating countries. This means that each country's airlines are allowed to fly between any city in their home country and any city in the participating countries. This type of agreement is intended to promote competition and new services through cooperative arrangements among the participating countries' airlines.

An interesting side of "Open Skies" is the increasing pressure on the United States to grant cabotage.

International Civil Aviation Organization (ICAO)

Origins

ICAO origins arise from the Chicago Conference. Although the economic issues discussed met with mixed success, the technical issues raised resulted in the Convention on International Civil Aviation. The implementing organization is ICAO, a specialized agency of the United Nations. It works with other UN members such as the World Meteorological Organization, the International Telecommunication Union, the Universal Postal Union, the World Health Organization, and the International Maritime Organization.

Nongovernmental organizations also participate in ICAO's work and include the International Air Transport Association, the Airports Council International, the International Federation of Airline Pilots' Associations, and the International Council of Aircraft Owner and Pilot Associations.

Responsibilities of contracting states

The 96 articles of the Chicago Convention establish the obligations of member states. A brief summary of particularly significant articles is shown in Table 1-4. There are presently 185 contracting states that belong to the organization.

Mandate

One of ICAO's chief activities is the establishment of international standards, recommended practices, procedures, and guidance material. An example is Communications, Navigation, and Surveillance/Air Traffic Management (CNS/ATM), formerly known as the future air navigation systems (FANS). This is the application of satellites, computer database data links, and cockpit avionic systems for CNS/ATM.

Table 1-4
Responsibilities of ICAO Contracting States

Article	Responsibilities of Contracting States
1	Contracting states recognize that each state has complete and exclusive sovereignty over the airspace above its territory.
3	The Convention, including the articles and annexes, applies only to civil aircraft, and each state will require their state aircraft to operate with "due regard" for the safety of navigation of civil aircraft.
11	International air navigation laws and regulations of a contracting state relating to the operation and navigation of such aircraft while within its territory shall apply to the aircraft of all contracting states without distinction to nationality. These laws and regulations shall be complied with by such aircraft while entering, within, or departing from the territory of that state.
12	Each contracting state undertakes to adopt measures to ensure that every aircraft maneuvering over or within its territory, and every aircraft carrying its nationality mark, wherever it operates, shall comply with the rules and regulations of that country relating to the flight and maneuver of aircraft. This article also requires that over the high seas, the rules in force shall be those established under this convention. Each contracting state undertakes to ensure the prosecution of all persons violating the applicable regulations.
15	Each contracting state undertakes not to discriminate in the availability of, or changes for, airports and other air navigation facilities.
22	Each contracting state undertakes to provide in its territory, airports, radio services, meteorological services, and other air navigation facilities to facilitate international air navigation, in accordance with the standards and practices of ICAO.
28	Contracting states undertake to adopt and put into operation appropriate standard systems of communication procedures, codes, markings, signals, lighting, and other operational practices and rules recommended or established by ICAO.

Table 1-4
Responsibilities of ICAO Contracting States (*Continued*)

Article	Responsibilities of Contracting States
33	Contracting states recognize the validity of Certificates of Airworthiness and Licenses of Competency issued by other contracting states, when issued under conditions which comply with ICAO Standards.
37	Contracting states should collaborate in securing the highest practicable degree of uniformity in regulations, standards, procedures, and organization in relation to aircraft, personnel, airways, and auxiliary services in all matters when uniformity will facilitate and improve air navigation.
38	Each contracting state undertakes to immediately notify ICAO of any differences between national regulations and any ICAO Standards.

To assist states (countries) in planning their air transport services ICAO collects and publishes world aviation statistical data, and conducts economic studies.

ICAO promotes civil aviation in developing countries. Assistance consists of advising on the organization of government civil aviation departments and on the location and operation of facilities and services, particularly in the recruitment and administration of experts, training, and procurement of equipment.

Civil aviation training centers are located in Egypt, India, Indonesia, Jordan, Kenya, Morocco, Nigeria, Pakistan, Thailand, and Tunisia.

Regional planning

ICAO is involved with many regional projects. Regional air navigation meetings consider the requirements of air operations in specified areas. Facilities and services and creation of supplementary procedures necessary to support such items as the increases in traffic density, new air routes, and the introduction of new types of aircraft are examples.

Air Navigation Plans provide details of facilities, services, and procedures required for international air navigation within specified areas. Each plan contains recommendations for navigation facilities and services within a specific region.

Organization

The United Nations recognizes ICAO as a specialized agency for international civil aviation. An agreement between these organizations is designed to ensure an efficient working relationship and a mutual recognition of their respective roles. ICAO is not subordinate to, and does not receive any line of command authority from, the United Nations.

Assembly

The assembly, the governing body of the organization, meets every 3 years for a detailed review of the organization's technical, economic, legal, and technical assistance programs, and offers guidance concerning the future work of other ICAO bodies. Each nation has one vote in the assembly and unless the convention provides otherwise, a majority rules.

ICAO Council

The council is composed of elected representatives from 33 member states. It investigates situations that might create obstacles to international air navigation, and takes action as necessary to protect global air safety and order. When required, it also serves as an arbiter between member states on aviation matters.

Air Navigation Commission

The Air Navigation Commission is composed of 15 individuals, each considered an expert in a technical field of aviation. The group is concerned with the development of ICAO Standards and Recommended Practices.

Air Transport Committee

The Air Transport Committee's prime concern is for economic matters relating to airports, route facilities, and air carrier tariffs. This information is used to promote fair and equal opportunities.

Joint Support Committee

The Joint Support Committee provides for financial arrangements of air facilities or services when member states have inadequate resources. Most funding comes from direct user charges to air carriers. This committee studies air service problems and makes suitable arrangements between user and provider states.

Legal Committee

The Legal Committee interprets questions on the Chicago Convention and public and private law. Some of its main concerns include hijackings and other acts of air terrorism, air carrier liability, and jurisdiction over offenses on international flights.

Within the more than 180 contracting states of ICAO, there are many legal philosophies and many different systems of jurisprudence. The organization facilitates the adoption of international air law instruments and promotes their general acceptance. Air law instruments that have been adopted include:

- The international recognition of property rights in aircraft: damage done by aircraft to third parties on the surface
- The liability of the air carrier to its passengers: crimes committed on board aircraft
- The marking of plastic explosives for detection
- Unlawful interference with civil aviation

Unlawful interference with international civil aviation

The Committee on Unlawful Interference with International Civil Aviation and Its Facilities assists and advises the council on all activities relating to aviation security.

Secretariat

The secretariat headed by a council-appointed secretary general provides for ICAO's daily working needs. It consists of several bureaus providing the day-to-day development of technical and economic data.

ICAO annexes

Annexes to the convention identify broad standards and recommended practices. Standards are directives that ICAO members agree to follow. If a member state has a standard different from an ICAO Standard, that member must notify ICAO of the difference. Recommended practices are desirable practices but not essential. The annexes are supplemented by numerous ICAO publications that detail implementation. The 18 annexes are shown in Table 1-5.

Foreign legislative and regulatory processes

Most countries have legislative processes and regulations. By international agreement, signatories to the Chicago Convention agree to formulate their legislation and regulation to the recommendations of ICAO.

Some countries adopt the regulations of other countries like the United States or the United Kingdom. This may include contracting with other governments for certification and oversight activities as well. As an example, many of the countries within the French community (former colonies) in West Africa operate under French rules and contract with the Directorate General of Civil Aviation (DGCA) inspection and oversight.

The joint civil aviation authorities

In 1970, 13 European countries formed the Joint Airworthiness Authority (JAA). The purpose is to produce common European certification regulations.

The stated JAA objectives are to:

- Ensure common high levels of aviation safety within member states including combining the expertise of several countries to conduct Type Certification assessments.
- Contribute to fair and equal competition within member states by removing the hidden trade barriers from the technical regulations.
- Aim for cost-effective safety and minimum regulatory burdens so as to enhance European international competitiveness.

By 1987, 12 of the countries signed a Memorandum of Understanding (MOU), which committed them to complete all airworthiness codes and adopt them as their only standard. Presently the JAA system consists of members from 27 European national aviation authorities.

JAA has no control over resources, as each national authority has its own funding system. It is principally funded by the member countries, 20 percent of which comes from the United Kingdom, France, and Germany.

Harmonization

The FAA, JAA, and the industry work to eliminate significant differences in regulations including their interpretation and application. They participate in each other's rulemaking process with a view to harmonized or at least similar sets of regulations. Harmonization activity by the FAA is under the umbrella of an ARAC committee. JAA regulations are known as Joint Aviation Regulations (JAR). Their organization is similar to United States FAR.

Table 1-5
ICAO Annexes

Annex	Title	Subject
1	Personnel Licensing	Provides information on licensing of flight crews, air traffic controllers, and aircraft maintenance personnel.
2	Rules of the Air	Contains rules relating to conducting visual and instrument flight.
3	Meteorological Service for International Air Navigation	Provides for meteorological services for international air navigation and reporting of meteorological observations from aircraft.
4	Aeronautical Charts	Contains specifications for aeronautical charts used in international aviation.
5	Measurement Units Used in Air and Ground Operations	Lists dimensional systems to be used in air and ground operations.
6	Operation of Aircraft	Enumerates specifications that ensure a level of safety above a prescribed minimum in similar operations throughout the world.
7	Aircraft Nationality and Registration Marks	Specifies requirements for registration and identification of aircraft
8	Airworthiness of Aircraft	Specifies uniform procedures for certification and inspection of aircraft.
9	Facilitation	Provides for simplification of border-crossing formalities
10	Aeronautical Telecommunications	Provides for standardization of communications equipment, systems, and communications procedures.

11	Air Traffic Services	Includes information on establishing and operating air traffic control, flight information, and alerting services.
12	Search and Rescue	Provides information on organization and operation of facilities and services necessary for search and rescue.
13	Aircraft Accident Investigation	Provides for uniformity in notification, investigation, and reporting on aircraft accidents.
14	Aerodromes	Contains specifications for the design and equipment of aerodromes.
15	Aeronautical Information Services	Includes methods for collecting and disseminating aeronautical information required for flight operations.
16	Environmental Protection	Contains specifications for aircraft noise certification, noise monitoring, and noise exposure units for land use planning and aircraft engine emissions.
17	Security—Safeguarding International Civil Aviation Against Acts of Unlawful Interference	Specifies methods for safeguarding international civil aviation against unlawful acts of interference.
18	The Safe Transport of Dangerous Goods by Air	Contains specifications for labeling, packing, and shipping dangerous cargo.

European Aviation Safety Authority (EASA)

Liberalization of air transport in Europe has become common national policy through the actions of the European Union (EU).[26] But safety and safety monitoring remain fragmented. The safety and standardization efforts of the JAA are difficult to translate into national law because its standards are not compulsory and often follow national interests. National authorities are free to adopt or reject their resolutions. JAA is not a legally recognized body with treaty granted authority. It lacks political authority to conduct international activities. Rather it is a coordinator, information clearing house, and consensus builder for developing harmonized standards in much of Europe.[27] It is slow.

The European Union opened negotiations in the late 1990s with the parties of the JAA, including those not part of the EU. The intent is to establish a single organization responsible for civil aviation safety (EASA). The vision is to create the equivalent of a European "FAA" that has legal authority among the member countries.

[26]The 15 EU member states include Austria, Belgium, Denmark, Finland, France, Germany, Greece, Ireland, Italy, Luxembourg, The Netherlands, Portugal, Spain, Sweden, and the United Kingdom.

[27]Where binding, international agreements in such areas as airworthiness and aviation safety are concerned, the FAA must work on a bilateral basis with individual European aviation authorities because of the lack of legal standing of the JAA.

2

Controlling Agencies

Department of Transportation (DOT)

The Department is administered by the Secretary of Transportation. Among the Secretary's many duties is the regulation of air commerce. This includes promoting the development of a national air transportation system. It means responsibility for technical and economic regulation[1] of air carriers.

Anyone desirous of providing air service as an air carrier must obtain two separate authorizations from the department:

- Operating authority in the form of an Air Carrier Certificate[2] issued by the Federal Aviation Administration
- Economic authority[3] issued by the Office of Aviation and International Affairs in the form of a Certificate of Public Convenience and Necessity

[1]The Airline Deregulation Act of 1978 did not completely remove all aspects of economic regulation. The demise of the Civil Aeronautics Board saw many of the economic regulations that survived deregulation transferred to the Secretary. The act directed that the Civil Aeronautics Board be abolished on December 31, 1984. On January 1, 1985, remaining CAB functions were transferred to the Office of the Secretary of Transportation. International routes are still subject to economic regulation. Deregulation applies a standard of "fit, willing, and able." It provides open entry into a market without tariff setting. Domestic air carrier operations are deregulated.

[2]See Chapter 3 for the discussion of air carrier certification.

[3]It is extracurricular to this text to discuss the economic aspects of air carrier operations. Essentially regulation carried the concept of operations within the "public convenience and necessity" standard used by the CAB before deregulation. The standard restricted entry into markets and controlled fares.

Office of Aviation and International Affairs

This office

- Promotes United States interests in international aviation markets.
- Licenses U.S. and foreign carriers to serve in international air transportation.
- Supports multilateral and bilateral aviation negotiations with foreign governments. Working in cooperation with the U.S. State Department it participates on the U.S. negotiating delegations.
- Processes and resolves complaints concerning unfair competitive practices in international fares and rates.
- Determines the disposition of requests for approval and immunity from the antitrust laws of international aviation agreements.

Concerning domestic aviation the office is charged with economic regulation of United States air carriers. This includes:

- Issuance of certificates of public convenience and necessity.[4] Certificates are issued for
- Interstate[5] or foreign carriage[6] of passengers and/or cargo and mail authority
- All-cargo air transportation
- Authorization of commuter air carriers
- Establishing international and intra-Alaska mail rates
- Administration of the essential air service program

[4]Air carriers must hold economic certificates or an exemption in order to provide air transportation to the public.

[5]Note that the Department's authority for economic regulation does not apply to intrastate operations. Intrastate operators, however, are frequently regulated by individual state public utilities commissions.

[6]There are several types of "carriage." See Appendix, Definitions, for these. Frequently the applicability of specific Federal Aviation Regulations is dependent upon the type of "carriage" being conducted by an air carrier.

The Federal Aviation Administration (FAA)

The Administrator

The Secretary of Transportation's duties include the appointment of an Administrator to carry out the duties of the Secretary related to aviation safety.[7]

Mandate

The Administrator issues and enforces rules, regulations, and minimum standards relating to a large area of aeronautical activities. Its authority is contained in USC Title 49.[8] A high-level organization chart of the Administration is shown in Figure 2-1. The organization is multifaceted. Its activities include several principal areas.

Airworthiness

The Administration establishes and oversees the enforcement of standards for the design, manufacture, operation, and maintenance of aircraft and manned free balloons, propellers, powerplants, and accessories.

Air carrier technical regulation

With respect to air carrier operations, the role of the FAA is the regulation of an airline's core technical activities, principally, flight operations and maintenance. It is not to regulate the airline as a business.

Airmen certification and rating

Certification and rating (including medical) of airmen. This includes pilots, flight engineers, navigators, mechanics, repair personnel, dispatchers, ground instructors, and parachute riggers.

Air agency certification

Certification and surveillance of agencies such as repair stations, mechanic and pilot schools.

[7] "In the performance of the functions of the Administrator and the Administration, the Administrator is authorized to issue, rescind, and revise such regulations as are necessary to carry out those functions...." 49 USC 106(f)(3)(A).

[8] 49 USC Subtitle VII, Aviation Programs, defines the responsibilities of the FAA.

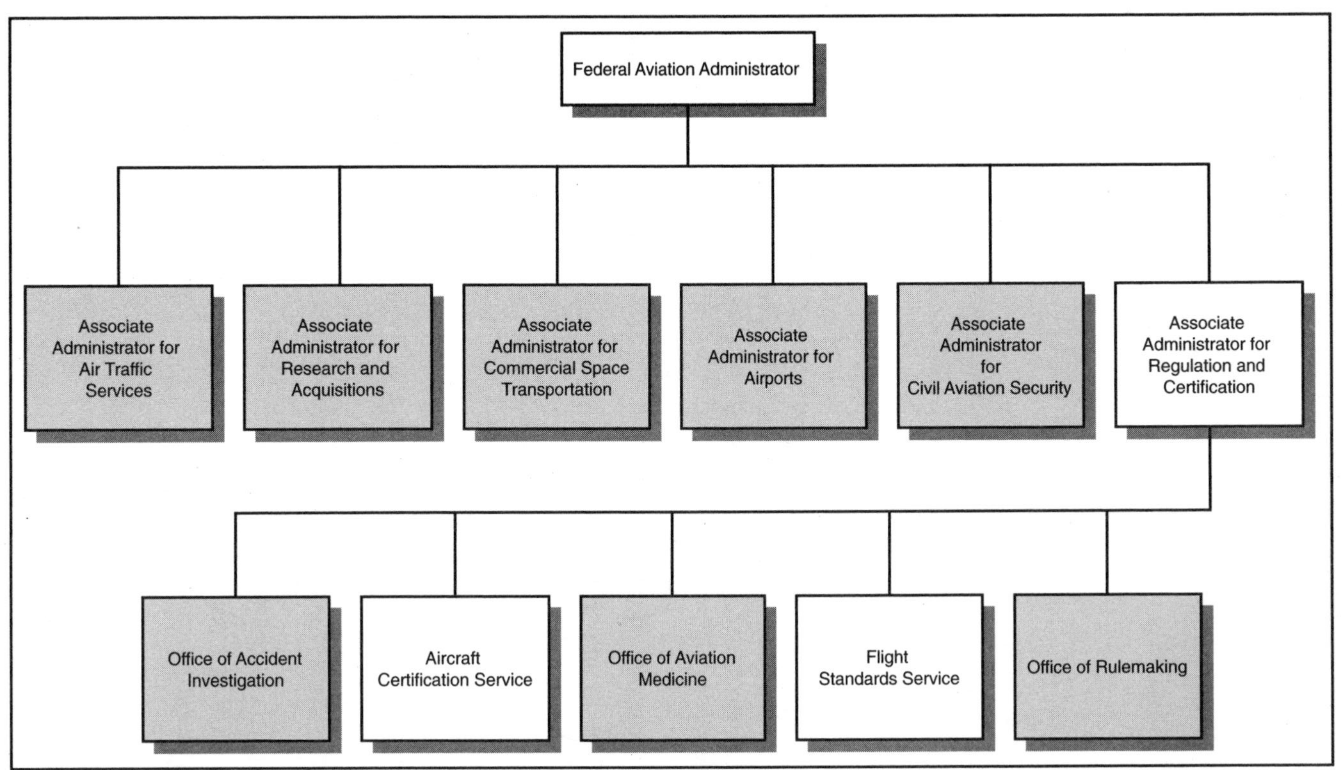

Fig. 2-1 *FAA organization.*

Registration and recordation

The Administration provides a system for the registration[9] of aircraft and recording of documents affecting title or interest in aircraft, aircraft engines, propellers, appliances, and spare parts.

Airport programs

The FAA maintains a national plan of airport requirements. It develops standards and technical guidance on airport planning, design, safety, and operations and provides grants to assist public agencies in airport system and master planning and airport development and improvement. It certifies airports serving air carriers.

Air navigation facilities

The Administration is responsible for federal visual and electronic aids to air navigation. It operates and maintains voice/data communications equipment, radar facilities, computer systems, and visual display equipment at flight service stations, airport traffic control towers, and air route traffic control centers. It performs flight inspections of air navigation facilities.

Airspace, air traffic management, and navigation

This consists of a network of airport traffic control towers, air route traffic control centers, and flight service stations. The system controls the use of the navigable airspace and the regulation of civil and military operations including common air traffic control.

It develops air traffic rules and regulations and allocates the use of the airspace. It also provides for the security control of air traffic to meet national defense requirements.

It develops specifications for the preparation of aeronautical charts. It publishes current information on airways and airport service.

Research, engineering, and development

The research, engineering, and development activities of the Administration provide the systems, procedures, facilities, and devices needed for air navigation and traffic control. The Administration supports development and testing of improved aircraft, engines, propellers, and appliances.

[9]49 USC Chapter 441, Registration and Recordation of Aircraft.

The Administration performs aeromedical research. It develops programs and regulations to control aircraft noise, sonic boom, and other environmental effects of civil aviation.

International activity

The Administration encourages aviation safety and civil aviation abroad. It exchanges aeronautical information with foreign aviation authorities; certifies foreign repair stations, airmen, and mechanics; negotiates bilateral airworthiness agreements to facilitate the import and export of aircraft and components; and provides technical assistance and training.

It provides technical representation at international conferences, including participation in the International Civil Aviation Organization and other international organizations.

Other programs

The Administration manages the aviation insurance program.[10] It issues technical publications for the improvement of safety in flight, airport planning, design, and other aeronautical activities.

The Administration regulates U.S. commercial space transportation that is within the sphere of the FAA. It licenses the private sector launching of space payloads on expendable launch vehicles and commercial space launch facilities. It also sets insurance requirements for the protection of persons and property and ensures that space transportation activities comply with U.S. domestic and foreign policy.

Hazardous materials

It enforces the hazardous materials regulations applicable to shipments by air as mandated in 49 CFR Part 107 and 49 CFR Part 171 through 175.

Antidrug programs

Since the late 1980s, the FAA has required specified aviation employers and operators to initiate antidrug programs,[11] including drug testing for personnel performing specified safety-related functions.

[10] 49 USC Chapter 443, Insurance.

[11] 49 USC, Chapter 451, Alcohol and Controlled Substances Testing, and 14 CFR Part 121, Appendix I, Drug Testing Program, and Appendix J, Alcohol Misuse Prevention Program.

Security programs

The FAA conducts research and development work regarding security, including antihijacking, and regulates air carrier and airport security[12] programs. It also provides assistance to law enforcement agencies in the enforcement of laws related to regulation of controlled substances.

Associate Administrator for Regulation and Certification

The FAA is diverse and huge. Our primary activity with the Administration may be found within two organizations of the Associate Administrator for Regulation and Certification: the Aircraft Certification Service and Flight Standards Service.

Aircraft Certification Service

The Aircraft Certification Service administers type certification and other airworthiness activities to determine compliance products with their basis of certification. This includes responsibility for defining continuing airworthiness.

Within this organization are four certification directorates (Figure 2-2). These are:

- Transport Airplane Directorate with headquarters in the Northwest Mountain region. This group is responsible for airplanes normally certificated directed by 14 CFR Part 25.
- Engine and Propeller Directorate (New England region). This group is responsible for the certification of power plants under 14 CFR Parts 33 and 35.
- Small Airplane Directorate with headquarters in the Central region. This group is responsible for airplanes normally certificated directed by 14 CFR Part 23.
- Rotorcraft Directorate (Southwest region). This organization handles the certification of helicopters certified under 14 CFR Part 27.

Within each directorate are one or more Aircraft Certification Offices (ACO). ACOs do the actual certification of aircraft and products. They work directly with the certificate applicants during certification.

[12] 49 USC 40119, 49 USC Chapter 449, Security, and 14 CFR Parts 107 and 108.

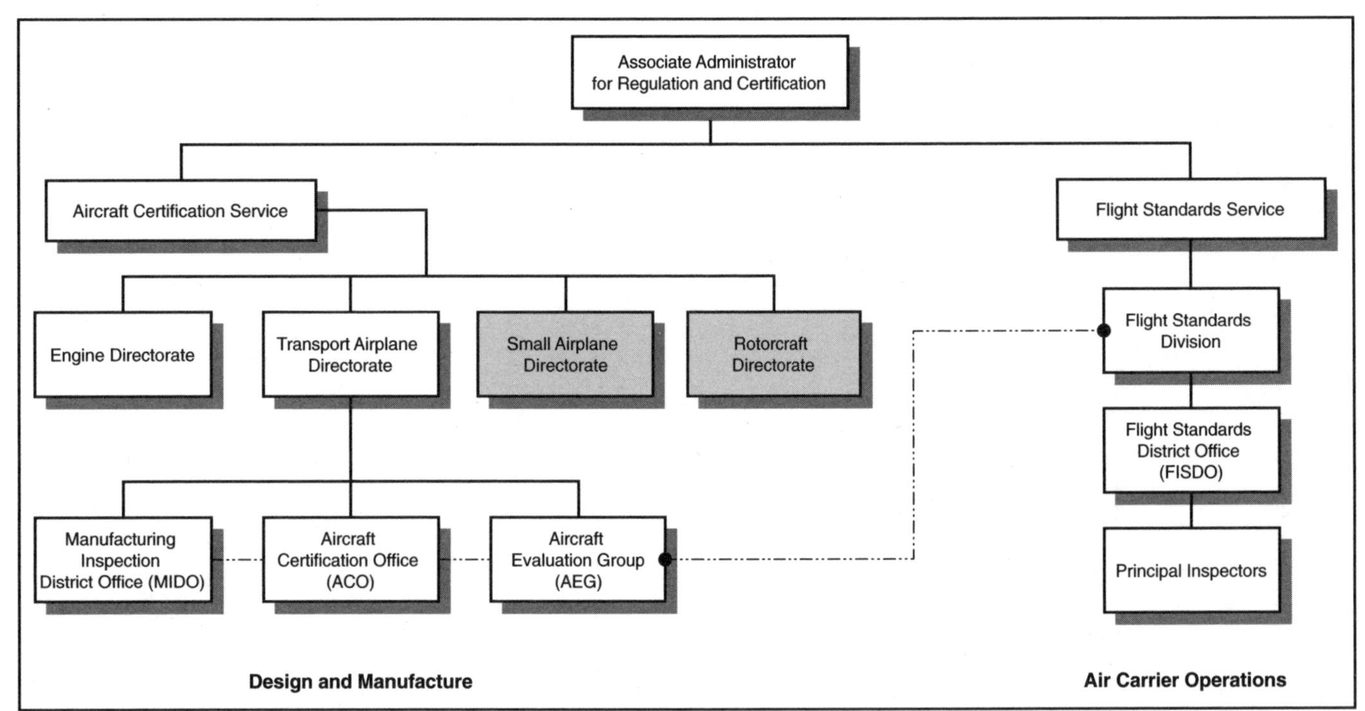

Fig. 2-2 *Aircraft Certification and Flight Standards Service.*

An Aircraft Evaluation Group (AEG), colocated with each directorate, is responsible to the Flight Standards Aircraft Evaluation Program staff manager for determining operational acceptability and continuing airworthiness requirements of newly certified or modified aircraft.

Flight Standards Service

Flight Standards Service is the organization (Figure 2-2) whose principal concern is with the operation and maintenance of aeronautical products.

Within the Washington headquarters of the service, there are two divisions that are our principal interest: Air Transportation Division (AFS-200) and the Aircraft Maintenance Division (AFS-300). They establish policy and procedures for administration by the Regional Flight Standard District (RFSD) offices.

Regional Flight Standards District (RFSD) offices, located in each of the nine FAA operating regions are responsible for managing and executing the daily certification, oversight, and enforcement functions through a system of Flight Standard District Offices (FSDO),[13] Certificate Management Offices (CMO), and International Field Offices (IFO).[14]

Principal inspectors

Operating out of the local FSDO is a group of safety inspectors. Within this group are the inspectors who have direct surveillance of certificated air carriers or air agencies such as repair stations, schools, and so forth.

In the air carrier world, we will have a minimum of two inspectors assigned to us, the principal operations inspector (POI) and the principal maintenance inspector (PMI). These two individuals are the primary FAA contact for a certificate holder. They are the first points of entry when dealing with the FAA. This begins during the air carrier certification process discussed in Chapter 3.

They have primary oversight responsibility for their assigned air carrier certificates. Inspectors may have several certificates for which

[13]There are approximately 100 FSDOs and satellite offices.

[14]International Field Offices (IFO) currently are located in Brussels, London, Frankfurt, and Singapore.

they are responsible. However, those inspectors assigned to a large airline usually have responsibility for one certificate. In fact, they may have additional safety inspectors who assist them.

Representatives of the Administrator

It is fundamental that within any given technical discipline there exist a finite number of individuals with suitable credentials and experience. The FAA, like any organization, must call upon this resource regardless of the availability of those individuals in the public or private sector. The law[15] recognizes this availability constraint and commissions the Administrator to use qualified private individuals (designees) to perform FAA examinations, testing, and inspections necessary to determine compliance with applicable standards.[16] The system assures the Administrator access to contemporary technology.

All designees work under the supervision of FAA personnel. They may be viewed as being FAA employees when exercising the privileges of their designation.

Each designation contains specific privileges and limitations. The number and type of designee is determined by need within the FAA organization. For example, an impending type certification project for a large aircraft may create a need for an increased number of designated engineering representatives. Designations thus may be withdrawn when the need no longer exists.

There are several categories of representatives:

- Designated engineering representatives (DER) approve engineering data within their discipline and the authority of their designation. Designees offer technical expertise with state-of-the-art knowledge. DERs are predominantly found at manufacturers where large amounts of engineering design data require review. Large repair stations will occasionally have DERs on the staff. Principal disciplines include:
  - Structural engineering.
  - Power plant engineering.
  - Systems and equipment engineering.

[15]49 USC 44702(d).

[16]14 CFR Part 183 *Representatives of the Administrator* contains the regulations governing FAA designees.

- Radio engineering.
- Engine engineering.
- Propeller engineering.
- Flight analyst.
- Flight test pilot.
- Acoustical engineering.

- Designated manufacturing inspection representatives (DMIR) are a part of the production process at manufacturer's facilities. They issue:
    - Original airworthiness certificates for aircraft and airworthiness approvals for engines, propellers, and product parts.
    - Export certificates of airworthiness and airworthiness approval tags.
    - Experimental certificates for aircraft for which the manufacturer holds the type certificate and which have undergone changes to the type design requiring a flight test.
    - Special flight permits to export aircraft.
- They conduct any inspections that may be necessary to determine that
    - Prototype products and related parts conform to design specifications.
    - Production products and related parts conform to the approved type of design and are in condition for safe operation.
- Designated airworthiness representatives (DAR) perform examination, inspection, and testing services necessary to issuing certificates in the areas of maintenance, manufacturing, and engineering.
- Designated aircraft maintenance inspectors (DAMI) approve maintenance on civil aircraft used by U.S. military flying clubs in foreign countries.
- Aviation medical examiners conduct medical examinations necessary to issuing medical certificates.
- Pilot examiners conduct oral and flight tests necessary to certify pilots.

- Technical personnel examiners conduct oral and practical tests necessary for issuing certificates to airman other than pilots. They include:
  - Designated mechanic examiner (DME).
  - Designated parachute rigger examiner (DPRE).
  - Air traffic control tower operator examiner.
  - Designated flight engineer examiner (DFEE).
  - Designated flight navigator examiner (DFNE).
  - Designated aircraft dispatcher examiner (DADE).

Authority of FAA over foreign air carriers

Foreign air carriers operating into the United States are subject under the Chicago Convention agreements to the operating regulations and air traffic management of the United States.

The FAA does not issue certificates to foreign air carriers. However, foreign air carriers operating to the United States are issued operations specifications.[17]

With respect to foreign operators, the dominant operating standards are found in 14 CFR Parts 91 and 129. The FAA's authority to regulate foreign operators using foreign registered aircraft is limited to assuring compliance with all the applicable operating rules, the ability to safely navigate and communicate within the U.S. national airspace system, and protecting persons and property on the ground. If a foreign operator uses any U.S. registered aircraft, other parts apply.[18]

Any U.S.-registered aircraft operated by a foreign air carrier is considered a foreign civil aircraft. The U.S. registration does not convey any additional authorization to the foreign air carrier beyond that granted in its operations specifications. The scope of the foreign air carrier's operations is limited to that granted in its DOT economic authority and reflected in its operations specifications issued by the FAA.

International law also makes it clear that the United States may require foreign operators to meet only the ICAO "standards," not the "recommended practices" of ICAO.

[17]14 CFR Part 129.11.

[18]For example 14 CFR Parts 21, 43, 61, 63, and 65.

FAA regulatory documents

Advisory Circulars (AC)

Advisory Circulars provide guidance and suggested methods for complying with a related Federal Aviation Regulation. Unless incorporated into a regulation by reference, Advisory Circulars (Figure 2-3) contain nonregulatory materials. Their contents, therefore, are not binding and are not considered the only means of compliance. Frequently the

Advisory Circular

Subject: USE OF CD-ROM SYSTEMS

Date: 8/14/97
Initiated by: AFS-350

AC NO: 120-69
Change:

1. PURPOSE. This advisory circular (AC) provides guidance on the use of CD-ROM (compact disk read-only memory) systems for the preservation and retention of the maintenance portion of a certificate holder's manual. The AC also provides guidance on the use of CD-ROM systems for the retrieval of the technical data contained in a certificate holder's manual.

2. FOCUS. This AC applies to certificate holders conducting operations under Title 14 of the Code of Federal Regulations (14 CFR) parts 121, 129, and 135. Operators under 14 CFR part 91 or 125 and repair stations certificated under 14 CFR part 145 also may use the guidance contained in this AC to the extent that it is applicable to the conduct of their operations.

3. RELATED MATERIAL.

a. 14 CFR §§ 43.13, 121.133, 121.135, 121.137, 121.139, 121.141, 121.369, 125.71, 125.73, 125.75, 125.249, 129.14, 135.21, 135.23, and 135.427.

b. AC 21-33, Quality Assurance of Software Used in Aircraft or Related Products; AC 21-35, Computer Generated/Stored Records Copies of these documents may be obtained from the U.S. Department of Transportation, Subsequent Distribution Office, Ardmore East Business Center, 3341 Q 75th Ave., Landover MD 20785.

c. Federal Aviation Administration (FAA) Order 8300.10, Airworthiness Inspector's Handbook; FAA Order 8400.10, Air Transportation Operations Inspector's Handbook; FAA Order 8700.1, General Aviation Inspector's Handbook. Copies of these documents may be purchased from: New Orders, Superintendent of Documents, P.O. Box 371954, Pittsburgh, PA 15250-7954.

FAA Form 1320-15 (4-82) Supersedes WA Form 1320-2

Fig. 2-3 *FAA Advisory Circular.*

accusation is made that the FAA circumvents the rulemaking provisions of the law by establishing "recommendations" in an AC. In practice, many circulars have become de facto regulations.

Circulars fulfill additional functions. Some are textbooks, informational pamphlets, or announcements of the availability of other publications.

Circulars are issued on an as needed basis. Drafts of significant or potentially sensitive circulars are published in the *Federal Register.* Public comment is solicited.

Civil Aeronautics Manuals

After the 1958 creation of the Federal Aviation Agency, the existing Civil Air Regulations (CAR) were updated and recodified into the existing FAR.

The CAR were supplemented by appropriately numbered Civil Aeronautics Manuals (CAM) that contained advisory policies, detailed technical information, procedures, and interpretations of Civil Aviation Regulations. For example, the old CAM 4b contained CAR 4b regulations and advisory materials for certification of transport category airplanes. CAMs were replaced by the existing system of FARs and associated Advisory Circulars.

Some CARs and CAMs remain applicable. The policies accompanying the CAR applicable to aeronautical products certified under the CAR continue to apply. The Boeing 727 was certified under CAR 4b, Airplane Airworthiness Transport Category. That regulation and the CAM continue to apply to that airplane.

Unfortunately CARs and CAMs are out of print and are difficult to locate.

FAA orders

These usually take the form of handbooks, which outline practices, policy, and procedures for various organizations within the FAA. Orders cover a wide range of subjects from how to issue AD notes to type certification procedures. Of particular relevance to the maintenance community is FAA Order 8300.10, Airworthiness Inspector's Handbook. The handbook contains operating procedures and policy. It includes a handbook bulletin system. Bulletins serve the function of alerting the field inspector on information that may require immediate action, special emphasis programs, or concerns that are time critical or temporary.

The FAA has broad discretion to interpret their own rules. Although what is in the handbooks doesn't have the force of law, the data in them can be used to cite violations.

Legal opinions

Legal opinions are issued by the FAA chief counsel's office in response to questions from individuals or agencies regarding the meaning of a given rule.

Policy statements

Federal agencies publish, in the *Federal Register,* general statements of policy[19] concerning internal processes and procedures. A general statement of policy may be issued for different purposes, including:

- Advice to the public on the manner in which the agency will exercise a discretionary power in subsequent adjudication or through rulemaking.
- Guidance to agency officials when exercising their discretionary powers (and, at the same time, notify the public of this guidance).

A general statement of policy may be issued to different audiences. Often policy statements address agency personnel, and sometimes they address the public. **A general statement of policy is "non-binding." This means that it does not constitute a new regulation. The FAA cannot apply or rely upon it as law.**

Unless the issuing agency specifically states that the policy is "proposed" and requests comments from the public to help develop the final policy, a general statement of policy may be considered effective when it is published in the *Federal Register.* It may have been in effect within the agency even before publication.

The National Transportation Safety Board (NTSB)

NTSB origins

The Air Commerce Act of 1926 established the first technical regulation of aviation outside the Post Office's interests. Contained within

[19] 65 FR 19959. FR Doc. 00-9214.

this legislation the Department of Commerce was given authority to determine the probable cause of airplane accidents. A small group within the department known as the Aeronautics Branch performed this function. It was a source of controversy from the beginning. The branch, uncomfortable in this role, shrouded its investigations in secrecy, an activity increasingly unpopular with the public.

An amendment to the Air Commerce Act in 1933 required the Secretary of Commerce to make public a report on the probable cause of each fatal aircraft accident. It gave the Secretary the power to subpoena witnesses and evidence in proceedings. It prohibited the accident report or other material gathered in an investigation from being used in civil or criminal proceedings. However, the legislation did not correct an apparent conflict of interest within the Aeronautics Branch.

By 1933, the Aeronautics Branch had become the Bureau of Air Commerce. The Independent Air Safety Board replaced this in turn. In 1938, the Air Safety Board was replaced by the newly formed Civil Aeronautics Administration (CAA). Two years later the Civil Aeronautics Board (CAB) was formed out of the CAA and charged with economic regulation of air commerce. Coincidental to the creation of the CAB, the Bureau of Safety was formed within the CAB and was charged with the accident investigative functions formerly held by the CAA.

The Bureau of Safety continued until 1966. At that time, the National Transportation Safety Board (NTSB) was established by the Department of Transportation Act of 1966. Although an independent agency it was made a part of DOT. The board was given the former CAB functions, powers, and duties concerning aviation accident investigations, formulating the probable cause of accidents, and making aviation safety improvement recommendations. It was further empowered by Congress to include accident investigations into four other transportation modes: rail, highway, marine, and pipeline. This discussion will limit itself to the aviation functions of the NTSB.

Congress passed the Independent Safety Board Act in 1974. This severed the association of the board with the DOT and gave to it increased investigative authority. By becoming independent of DOT, the NTSB was put in a more objective position for handling evaluations of DOT and FAA actions and officials, and for formulating safety recommendations.

The National Transportation Safety Board is responsible for transportation accident investigation in the United States. Since its inception in 1967, the NTSB has investigated more than 100,000 aviation accidents. The NTSB has issued almost 10,000 recommendations in all transportation modes. More than 80 percent of its recommendations have been adopted.

Mandate

It is important to note that the NTSB[20] is not a regulatory agency. It cannot fund or be directly involved in the operation of any mode of transportation. Rather its responsibility is the determination the "probable cause" of:

- All United States civil[21] and public use[22] government aircraft accidents
- Selected highway accidents
- Railroad accidents involving passenger trains or any train accident that results in at least one fatality or major property damage
- Major marine accidents and any marine accident involving a public and a nonpublic vessel
- Pipeline accidents involving fatality or substantial property damage
- Releases of hazardous materials in all forms of transportation
- Transportation accidents that involve problems of a recurring nature

Its function beyond the determination of probable cause is to offer recommendations to regulatory agencies regarding changes to regulation, policy, and practice it deems appropriate to correct identified transportation hazards.

Additionally the board

- Conducts special studies of safety problems.
- Maintains an official United States census of aviation accidents.

[20]The mandate of the NTSB is found in 49 USC Chapter 11.

[21]Note accidents involving military aircraft are not the board's responsibility.

[22]49 U.S.C. 40102(a)(37) defines public use aircraft. The definition is lengthy. Readers interested in specific details should consult the USC. Additional information may be found in FAA Advisory Circular AC 00-1.1, Government Aircraft Operations.

- Evaluates the effectiveness of government agencies involved in transportation safety.
- Evaluates the safeguards used in the transportation of hazardous materials.
- Evaluates the effectiveness of emergency responses to hazardous material accidents.
- Serves as the "court of appeals" for any airman, mechanic, or mariner whenever certificate action is taken by the Federal Aviation Administration or the U.S. Coast Guard commandant, or when civil penalties are assessed. This is done through the Office of Administrative Law Judges.
- Leads U.S. teams on foreign airline accident investigations to assist foreign authorities under the provisions of the International Civil Aviation Organization (ICAO) agreements.

To ensure that safety board investigations focus only on improving transportation safety, the board's analysis of information and its determination of probable cause cannot be entered as evidence in a court of law.

Organization

The safety board (Figure 2-4) is composed of five members nominated for 5-year terms by the President and confirmed by the Senate. Two of the members are designated by the President to serve as chairman and vice chairman for 2-year terms.

The board is supported by a staff of approximately 400 people representing several engineering and scientific disciplines under the direction of the managing director. Board headquarters is in Washington D.C. Regional offices (Figure 2-5) are located throughout the country.

The Office of Aviation Safety

The Office of Aviation Safety conducts investigations of all aviation accidents within the board's jurisdiction. It prepares reports for submission to the board and release to the public. These set forth the facts and circumstances of accidents including a recommendation as to the probable cause(s) and safety recommendations to prevent future aviation accidents.

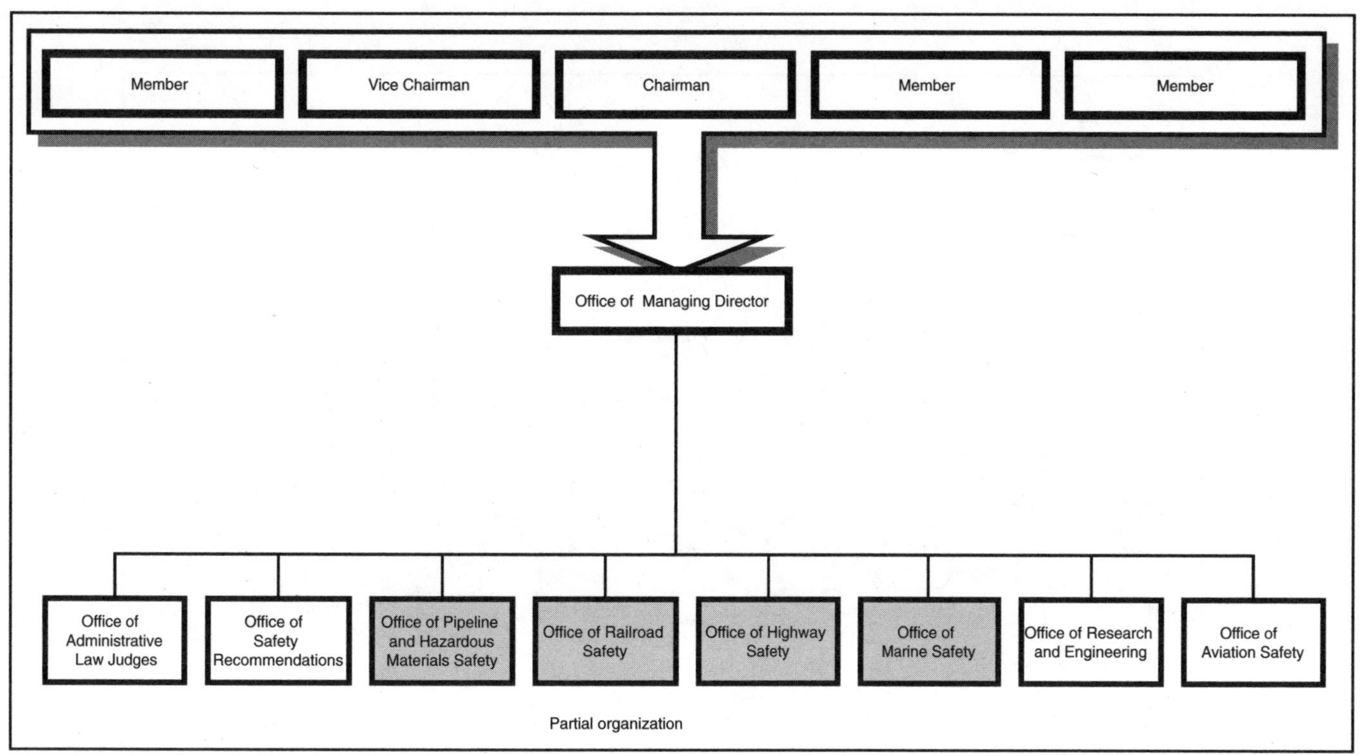

Fig. 2-4 *NTSB organization.*

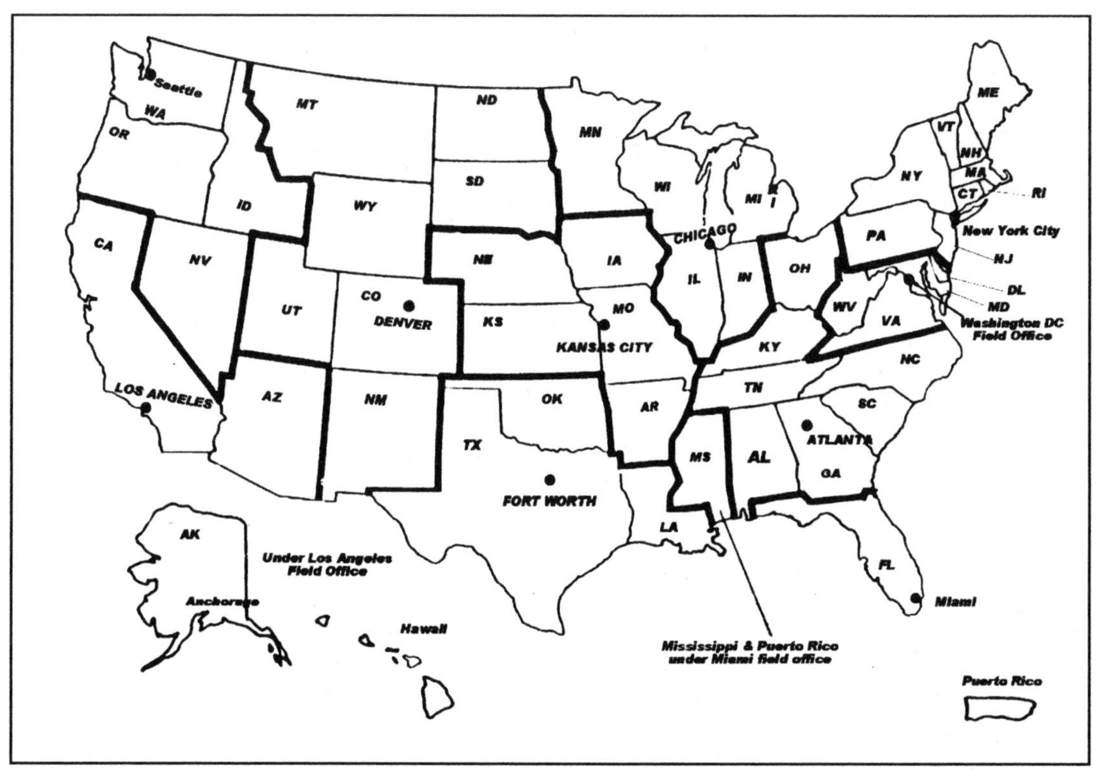

Fig. 2-5 *NTSB regional offices.*

The office participates in the investigation of accidents that occur in foreign countries and involve U.S.-registered and/or U.S.-manufactured aircraft. It conducts special investigations into selected aviation accidents involving safety issues of concern to the board.

The office consists of six divisions: Major Investigations, Operational Factors, Human Performance, Survival Factors, Aviation Engineering, and Regional and General Aviation.

The Office of Research and Engineering

The Office of Research and Engineering provides technical advice and services. It conducts research and carries out analytical studies and tests on all aspects of the board's accident investigation, accident prevention, and safety promotion activities.

The Office of Safety Recommendations

Safety recommendations are a part of the safety board's mandate. The Office of Safety Recommendations oversees recommendations.

The board must address safety deficiencies immediately. It often issues recommendations before the completion of investigations. These are based on findings developed during the conduct of an investigation. They may address deficiencies that do not pertain directly to what is ultimately determined to be the cause of the accident.[23]

As of February 1999, approximately 84 percent of the board's aviation recommendations have been accepted.[24]

The NTSB "Most Wanted" List

Beginning in 1990, the board has highlighted some issues on a NTSB Most Wanted list (Figure 2-6) of safety improvements. This program is intended to spotlight safety recommendations issued by the board that it feels will have the greatest impact on transportation safety.

[23]For example, in the course of its investigation of the crash landing of a DC-10 in Sioux City, Iowa, in 1989, the board issued recommendations on four separate occasions before issuance of its final report. In the case of the crash of an ATR-72 in Roselawn, Indiana, in 1994, the board issued urgent safety recommendations within 1 week of the accident. In the TWA flight 800 investigation, once it was determined that an explosion in the center fuel tank caused the breakup of the aircraft, the board issued urgent safety recommendations aimed at eliminating explosive fuel/air vapors in airliner fuel tanks.

[24]*Safety Recommendations Statistical Information,* NTSB report, 1999. It consists of tables and charts depicting statistical information. It is available on the NTSB website.

NTSB Most Wanted Transportation Safety Improvements

Item	Description	Action Needed by
Positive Train Separation	Require a railroad collision avoidance system.	The Federal Railroad Administration and the railroad industry
Recreational Boating Safety	Strengthen legislation, enforcement, and education programs to prevent boating accidents.	The states
Human Fatigue in Transportation Operations	Study the relationship of fatigue in the transportation industry and update applicable regulations.	The Department of Transportation (DOT), the Federal Aviation Administration (FAA), the Federal Highway Administration (FHWA), the Federal Railroad Administration (FRA), the Federal Transit Administration (FTA), the U.S. Coast Guard (USCG), the Research & Special Programs Administration (RSPA), the Brotherhood of Locomotive Engineers (BLE), the United Transportation Union (UTU), the Association of American Railroads (AAR), the American Public Transit Association (APTA), and the New York City Transit Authority (NYCTA).
Automatic Information Recording Devices	Require devices that will automatically record specified information.	The United States Coast Guard, the Federal Highway Administration, the National Highway Traffic Safety Administration, the Federal Aviation Administration, various trucking associations, the Teamsters Union, and the American Public Transit Association

Child/Youth Safety in Transportation	Enact graduated driver licensing legislation; educate public about transporting kids in back seats; make the back seat more child friendly; and require restraints for infants and small children on airplanes.	The States, the National Highway Traffic Safety Administration (NHTSA), the Federal Aviation Administration (FAA), the automobile industry, and the child restraint industry
Airframe Structural Icing	Revise icing criteria and certification testing requirements. Research and develop onboard aircraft ice protection and detection systems.	The Federal Aviation Administration
Excavation Damage Prevention to Underground Facilities	Require the installation of excess flow valves in high-pressure residential natural gas distribution systems and provide education and training related to third-party damage.	The Research and Special Programs Administration, pipeline industry, the states, the American Gas Association, the American Public Works Association, and the American Public Gas Association
Explosive Mixtures in Fuel Tanks on Transport Category Aircraft	Require preclusion of operation of transport category aircraft with explosive fuel-air mixture in fuel tanks.	The Federal Aviation Administration
Airport Runway Incursions	Provide for safer control of aircraft on the ground.	The Federal Aviation Administration
Highway Vehicle Occupant Protection	Enact primary enforcement of state seat belt laws and evaluate higher air bag deployment thresholds.	The National Highway Traffic Safety Administration, the states, and automobile manufacturers

Fig. 2-6 *NTSB Most Wanted list.*

The Office of Administrative Law Judges

Among its duties the Office of Administrative Law Judges is responsible for the conduct of all formal proceedings arising under the Federal Aviation Act of 1958. These primarily involve appeals by airmen, air agencies, and air carriers from FAA orders suspending, revoking, or modifying their certificates. Administrative law judges conduct objective reviews of evidence, develop records, apply judicial expertise, and render decisions.

Appeals arise from alleged violations of the Federal Aviation Regulations, or from lack of qualification to hold such certificates. Proceedings can also involve petitions from applicants denied airman medical or other certification by the FAA Administrator and certain civil penalty actions initiated by the FAA Administrator. Additionally, this office reviews and decides applications for attorney fees and expenses under the Equal Access to Justice Act from certificate holders who prevail against the FAA in certain adversary adjudication.

The board is not bound by FAA findings. Administrative Law Judges are consequently independent, impartial triers of fact in formal hearings similar to that of a trial judge conducting civil trials without a jury.

When an FAA order is appealed to the NTSB, the case is assigned an Administrative Law Judge for a hearing. An advance notice of formal hearing is issued at least 30 to 60 days before this hearing. Notices given in an emergency hearing can be in as little as 2 weeks because the board is required by law to dispose of these cases in 60 days. Hearings can last anywhere from several hours to several days. A typical hearing lasts from 1 to 2 days.

At formal hearings, the parties are given an opportunity to present oral and documentary evidence, to rebut, to conduct cross-examinations, and to make arguments. Counsel or other qualified representatives may represent the parties. The FAA bears the burden of proving the charges alleged in its orders. Petitioners may offer evidence in defense. When cases involve certificate denials, the petitioner bears the burden of proof.

In most cases, judges issue an initial oral decision at the conclusion of the hearing. If either the FAA or the petitioner is dissatisfied with the judge's decision, a further appeal may be taken to the full five-member board. If either party is dissatisfied with the full board's order, judicial review may be obtained in a federal appeals court.

However, the FAA can only appeal the board's order in cases that it determines may have a significant adverse impact on the implementation of the legislative mandates.

A certificate remains effective until the NTSB finally disposes of the appeal, except in cases where the FAA has determined an emergency in safety exists. In those cases, an Emergency Order of Revocation, Emergency Order of Suspension, or an Order with Immediate Effectiveness will be issued and the petitioner must surrender the certificate, pending the outcome of an appeal.

The accident investigation process

Accident investigation is a comprehensive process involving:

- On-site investigation and gathering of evidence
- Preliminary analysis of evidence
- Public hearings and taking of testimony
- Final analysis of evidence and testimony
- The preparation of a public report by the board including a statement of "The probable cause..."

The party system

Diverse organizations can be designated as parties to an investigation. The board has complete discretion over which organizations it designates "parties" to the investigation.[25] Only organizations that can provide expertise to the investigation are granted party status.

The use of the party system leverages the limited resources of the board, assuring technical expertise as well as increased physical resources. Persons in legal or litigation positions are not allowed party status to the investigation. All party members report to the NTSB.

The role of the FAA in accident investigation

The Federal Aviation Administration (FAA) participates actively in every aircraft accident investigation, having been directed to do so by Congress.[26] The FAA Administrator[27] designates a representative

[25]The FAA, by law, is automatically a designated party. See succeeding paragraphs.

[26]See 49 USC 1441(g).

[27]Frequently cited in legislation as the secretary of transportation.

to actively participate in every major aircraft accident investigation, and every full field investigation. The presence of the FAA provides additional technical expertise, as well as institutional knowledge.

On occasion, at the board's request, the FAA investigates certain aviation accidents.[28] The board is authorized to use the FAA report of an accident so delegated in making its determinations of probable cause. However, the FAA does not participate in determining "probable cause" findings.

Investigations involving criminal activity

In cases of suspected criminal activity, other agencies may participate in the investigation. The Safety Board does not investigate criminal activity. Once it has been established that a transportation tragedy is, in fact, a criminal act, the FBI becomes the lead federal investigative body. The NTSB provides requested support to criminal investigative agencies.

The investigation

An investigator in charge (IIC) conducts investigations. In a major investigation, the IIC may establish investigative working groups, made up of specialists from the parties. A board investigator, called the group chairman, leads each group (Figure 2-7). Only those persons who can provide the board with needed technical or specialized expertise are permitted to serve on a team. The number and composition of groups varies depending on the nature of the accident. Groups are commonly divided by discipline. Typical are aerodynamics, flying qualities, structures, systems, power plants, human performance, fire and explosion, meteorology, radar data, flight and voice data recorder, air traffic control, witness statements.

Eventually, each investigative group chairman prepares a factual report. Each of the parties in the group is asked to verify the accuracy of the report. The factual reports are placed in the public docket.

Public hearings

The board may hold a public hearing as part of an investigation. The purpose is

[28]Appendix O of Part 800 of the NTSB Organizational Rules sets forth a delegation of authority to the FAA to investigate certain selected kinds of accidents on behalf of the NTSB.

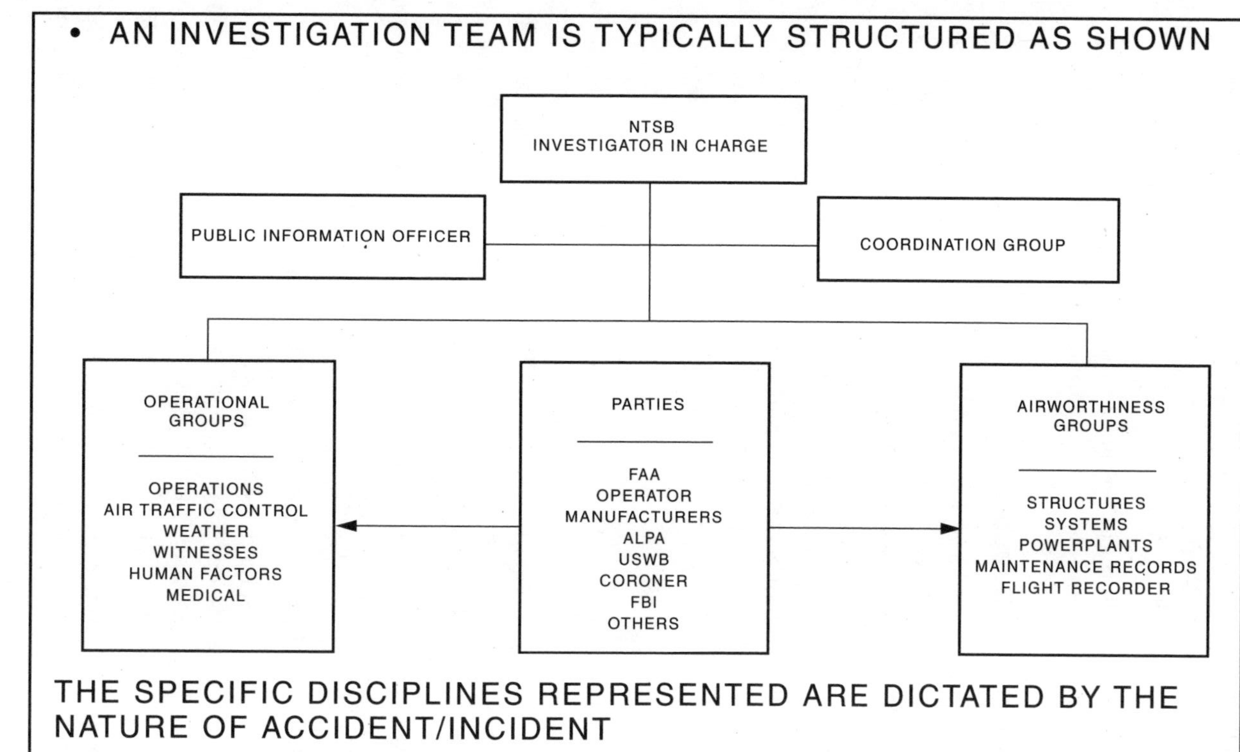

Fig. 2-7 *Accident Investigation Team Organization.*

- To gather sworn testimony from subpoenaed witnesses on issues identified by the board during the course of the investigation
- To allow the public to observe the progress of the investigation

Hearings are usually held within 6 months of an accident, but may be postponed for complex investigations.

Analysis and draft report

Eventually the safety board staff analyzes the group reports, physical evidence, testimony, and test results. A final report is drafted. The draft presents the evidence, its analysis and proposed probable cause and recommendations.

Parties to the investigation do not participate in the analysis and report writing phase of NTSB investigations. However, they are invited to submit their proposed findings of cause and safety recommendations. These are made part of the public docket.

Board meetings

The board members deliberate over the draft final report during a public meeting in Washington, D.C. The members evaluate the staff's draft. They may direct the staff to modify the report during these deliberations. Upon acceptance of the report by the five members, the report is published (Figure 2-8).

Non-safety board personnel, including parties and family members, cannot interact with the board during these meetings.

The NTSB role in international accident investigation

ICAO[29] Annex 13, Aircraft Accident Investigation, outlines the basic requirements and responsibilities for accident investigation. Responsibility for investigation of aircraft accidents and the issuance of findings is the responsibility of the member country in which the accident occurs.

The state of aircraft registry is responsible for investigation of accidents occurring in states that are not ICAO members, and the territory outside any state.

[29]See discussion in Chapter 1 for an explanation of the International Civil Aviation Organization (ICAO).

PB98-910401
NTSB/AAR-98/01
DCA96MA068

NATIONAL
TRANSPORTATION
SAFETY
BOARD

WASHINGTON, D.C. 20594

AIRCRAFT ACCIDENT REPORT

UNCONTAINED ENGINE FAILURE
DELTA AIR LINES FLIGHT 1288
MCDONNELL DOUGLAS MD-88, N927DA
PENSACOLA, FLORIDA
JULY 6, 1996

6725C

Fig. 2-8 *The accident report.*

Ancillary agencies

Many entities other than those just discussed affect the conduct of air carrier maintenance. These agencies are in addition to normally understood traditional agencies such as license/permit bureaus, public utilities commissions, taxing authorities, and labor unions. There are a surprising number of these ancillary agencies. The list seems endless.

The following discussions show some of these nontraditional business–related U.S. agencies and associations, which affect maintenance operations. It is by no means a complete summary.

They all must be considered within the context of international operations. There are available in each country into which an airline operates corresponding provincial, state, and municipal agencies whose requirements must be accommodated. It is essential when beginning operations or expanding into foreign countries that consideration be given to the influence these agencies have on maintenance operations.

Agricultural agencies

Airlines handle enormous quantities of agricultural products and live animals. Maintenance and engineering organizations are frequently charged with providing appropriate equipment for the handling and shipment of animals and produce. Cleanup, corrosion control, and fumigation are maintenance activities.

The U.S. Department of Agriculture enforces animal and plant imports and export regulations through agricultural quarantine inspections at border ports.

Various state agencies additionally control the import of agricultural products into their states. Excellent examples are the states of California and Hawaii. Both have a large agricultural industry and are thus very sensitive to protection against plant diseases and insect infestation.

Airport commissions/port authorities

Airports in the United States are predominantly publicly owned by municipalities or by municipal/state public corporations. An example of the latter is The Port Authority of New York and New Jersey. Revenue is intended to cover the operation/maintenance and development of the airport.[30]

[30]An interesting variation of the concept occurred in Los Angeles, California. The airport authorities in the early 1990s tripled the landing fees. The resulting revenue was diverted into the general tax fund. Los Angeles International Airport and the City of Los Angeles exhausted their last legal resort this week in a case of airport landing fee gouging, and must pay $100 million in refunds to the airlines. The U.S. Supreme Court declined to review an appeal from the city, which had tripled the landing fees at LAX in the early 1990s and diverted the cash to its general municipal fund. The city had already lost in the lower courts after a suit brought by the Air Transport Association.

Fees imposed on operators take on an almost infinite variety. Some are discussed below.

Landing fees

This is one of the principal sources of revenue for airports. Landing fees are usually based on the maximum certificated landing weight[31] of the airplane, as shown in the FAA Approved Airplane Flight Manual (AFM). The fee is usually expressed as a multiple of 1,000 pounds (or 1,000 kg) of the value. The fee is independent of the actual landing weight of the airplane. Thus, for example the certificated landing weight of the airplane may be 830,000 pounds. The fee may be expressed as $2.50 per 1,000 pounds. Thus, the landing fee is $2,075.

It should be noted that frequently the fee is assessed every time the wheels touch the ground. Thus conducting pilot flight training at a given airport may become very costly. Often an airline will request the manufacturer of the airplane to certificate it at a weight lower than its maximum structural capability. This keeps the landing fees down.

Facilities charges

These can be represented by charges for the use of certain facilities at the airport. Immigration and health clearance facilities, basic terminal services, and agricultural clearance facilities to mention just a few. They can take the form of a flat fee (rare) or a head tax that may be based on the actual number of passengers passing through a facility. They may however be occasionally based upon the number of installed seats on the airplane.

Parking fees may be imposed. These may be graduated on the basis of location of the airplane on the airport. This is used as an incentive to keep certain ramp or terminal areas clear, for example.

Terminal facilities charges or leases are negotiated for office space, gate space, and passenger check-in. Charges may be imposed for baggage handling use of airport conveyor systems. A flat fee or a graduated fee on the basis of an hourly rate may be charged for terminal gates.

Aircraft ramp handling services

Ground handling involves activities that take place on the ramp, such as loading and unloading aircraft, cleaning and maintaining

[31]In some instances, the fee may be based on the maximum takeoff weight. Regardless of the value chosen by the airport authorities, the principle is the same.

aircraft and equipment, refueling, pushing back and towing aircraft, loading and unloading baggage, and catering. It is a significant cost element affecting airlines.

For many airports, handling services are held as a monopoly. They assess fees and grant exclusive concessions to independent contractors such as caterers, fueling companies, baggage handling, and ramp services companies. Airlines find this costly. They like the freedom to conduct such services themselves or freely contract for them by choosing among competing agents.

Communications agencies

Federal Communications Commission

The Federal Communications Commission (FCC) is an independent agency directly responsible to Congress. It was established by the Communications Act of 1934. Five commissioners direct the commission.

The commission is charged with regulating[32] interstate and international communications by radio, television, wire, satellite, and cable. FCC rules include sections addressing the maintenance, repair, and overhaul of radio equipment. This includes the licensing of personnel who perform such work.

The comparable international agency is the International Telecommunications Union.

RTCA[33]

RTCA, Inc., is a private, not-for-profit corporation that develops recommendations regarding communications, navigation, surveillance, and air traffic management. RTCA functions as a federal advisory committee. Its recommendations are used by the Federal Aviation Administration (FAA) and by the private sector. Many non-U.S. government and business organizations also belong to the RTCA.

Organized in 1935 as the Radio Technical Commission for Aeronautics, RTCA today includes over 200 aviation-related organizations. RTCA guides the development and use of aviation systems and technology in response to airspace user needs. Its products are developed by special committees staffed by volunteers from its membership. The

[32]FCC regulations are contained in 47 CFR, Telecommunications.

[33]From the RTCA website.

principal product of the organization is the development of operating, technical, and performance standards.

Because RTCA functions as a Federal Advisory Committee, formation of the new special committee, as well as all committee meetings are announced in the *Federal Register*. Committee meetings are open to all interested parties.

SITA

SITA, founded in 1949, is a communications company based in Geneva, Switzerland. It consists of approximately 700 members principally from the airline and related industries. It provides communications services primarily to the aviation industry. The company owns and operates an international (some 138 countries) voice and data network, which has over 120,000 user connections worldwide.

Aeronautical Radio Inc. (ARINC)

ARINC is a communications company. Originally formed in 1929 it provides services to the airline industry. The company is owned by the member airlines. The company began providing the airlines with national en route flight-following and weather information in 1930. In addition to its communications services the company also is involved in equipment development including systems engineering, development, and integration including avionics and other technical standards.

Department of Defense

A source of revenue for U.S. carriers is providing charter service for the military through contracts issued by the Air Mobility Command (AMC) of the United States Air Force. These include scheduled cargo flights, and troop, dependent, and prisoner flights. The contracts carry provisions for the military oversight and inspection of maintenance facilities and practices.

The Civil Reserve Air Fleet (CRAF) is administered by the United States Air Force. CRAF supports Department of Defense airlift requirements during emergencies when the need for airlift exceeds the capability of military aircraft. Airlines contractually pledge aircraft to the CRAF when needed. AMC controls the aircraft missions when CRAF is activated.[34] The air carriers, however, continue to operate and maintain the aircraft with their resources.

[34]The last activation of CRAF was during the Gulf War in the early 1990s.

As incentives for carriers to commit aircraft to the CRAF, AMC awards peacetime airlift contracts to CRAF participant airlines. Furthermore, participating airlines are subsidized for aircraft modifications to their aircraft to meet CRAF requirements. This includes compensation for performance penalties incurred by such modification.[35] Approximately 35 FAA certificated airlines and 657 aircraft are enrolled in the program.

CRAF activation is only used when Department of Defense needs civil augmentation of airlift. Carriers agree to make their aircraft ready for a CRAF mission within 24 to 48 hours of call-up.

Immigration agencies

These government agencies conduct immigration inspections of travelers entering individual countries as they arrive at officially designated ports of entry.

In the United States the Immigration Service, which is a part of the U.S. Department of Justice, provides this function. There are approximately 250 ports of entry in the United States, which include air, land, and sea locations.

Customs Service

Government customs agencies control the import of goods into a country. Incidental to this mandate they have control of an airplane and its contents upon its arrival at a port of entry. This includes access to the airplane.

Department of Commerce

The department through its export control authority can influence the movement of some equipment most notably spare parts.

Postal agencies

Postal agencies through their own rules and those recommended by the International Postal Union control the handling of mail including on and around the airplane. They can and do affect maintenance operations.

[35]As an example, the modification of an aircraft to install a large upper-deck cargo door and cargo handling system increases the airplane's weight. This in turn imposes payload range performance penalties.

Public health agencies

These agencies monitor entering vessels and passengers for communicable diseases. They establish, with the World Health Organization, immunization requirements. They regulate food handling and potable water sources and systems.[36] The disposition of garbage and waste removed from international aircraft[37] is a function of these agencies. For example, the United States Public Health Service establishes regulation for the design of galley equipment and waste systems. Health agencies frequently work with agricultural agencies. For example, many countries require that pet dogs and cats brought into a country be quarantined and examined at the first port of entry for evidence of disease that can be transmitted to humans.

In the United States the U.S. Public Health Service (USPHS) is the predominant health agency. The service participates in the activities of the World Health Organization as well as those of various state and municipal agencies.

Trade associations

International Air Transport Association (IATA)

IATA was founded in 1945 with 57 member airlines from 31 nations. Today its membership includes over 200 organizations. It represents the interests of its members in national/international regulatory and legislative matters. IATA also holds a unique charter regarding the rates and tariffs previously discussed under the Chicago Convention.

It establishes common standards and practices, which are of a noncompetitive technical character. This includes a wide array of interests such as flight operations and safety, engineering, maintenance and materiel, live animal handling, hazardous goods handling, airport operations, air traffic management, cargo, electronic data interchange, facilitation, passenger service, public relations, and security. It operates an extensive training function providing both technical and management instruction.

[36]These functions may be further monitored by state and municipal agencies as well. The Department of Defense also has oversight of food handling under terms of charter contracts with commercial operators.

[37]They work in cooperation with agricultural agencies to prevent the importation of agricultural pests as well as potential threats to public health.

Air Transport Association (ATA)

The ATA, established in 1936, is a trade organization for U.S. commercial airlines. Like IATA, it represents the interests of its members in regulatory and legislative matters. As a technical organization, it establishes common standards and practices that are of a noncompetitive technical character. The association deals with operations and safety, engineering, maintenance and materiel, airport operations, air traffic management, cargo, electronic data interchange, facilitation, federal and state government affairs, international affairs, legal affairs, passenger service, public relations, and security.

The National Weather Service

The National Weather Service is an agency within The National Oceanic and Atmospheric Administration (NOAA). It provides weather forecasting and reporting services to the aviation community. The National Weather Services international organizational relationships are principally associated with the World Meteorological Union.

At first glance, it would appear that the services are predominately of interest to flight operations. However, weather information is equally important to maintenance organizations. Weather affects aircraft handling, deicing, snow removal, and positioning. It requires that airplane dispersal plans be available for severe weather warnings.

3

Air Carrier Certification

Introduction

Individuals desiring to operate as air carriers require two certifications, economic and operating. The Department of Transportation (DOT) Office of the Secretary issues economic authority. Operating authority is the purview of the Federal Aviation Administration.

An airline needs three documents in order to operate: an Economic Certificate, Operating Certificate, and the Operations Specifications. The three are independent of each other. If any one of them is missing, operations may not be conducted.

Terms

You will see throughout the regulations, law, literature, and this chapter an overwhelming and confusing use of terms when referring to air carriers or the conduct of air carrier operations. These arise from words/phrases used to define various classes of air service or the conduct of a given type of flight operation. They are used to identify the specific applicability of certain provisions of the law or regulation. An air carrier may be classified under more than one of these terms. They are just names used to categorize air transportation activity. Definitions are in the Appendix.

Economic authority

The Secretary of Transportation issues a certificate of public convenience and necessity to a citizen of the United States[1] authorizing the

[1]Only a citizen of the United States may be an "air carrier."

provision of air transportation consisting of the following types of service:

- Interstate or foreign, passenger, and/or cargo authority issued under 41102 of USC Title 49
- All cargo authority under 41103 of USC Title 49

Certificates are issued when the Secretary determines that the applicant is "fit, willing and able" to perform the service.

When issuing a certificate, the Secretary may impose terms considered necessary to the public's interest. Authority may be granted for scheduled or charter operations serving domestic or foreign points, carrying passengers/cargo/mail or cargo/mail only. However, terms may not be imposed that restrict the places served or prices charged by the certificate holder.

Specific authority consists of the certificate itself together with any terms, conditions, or limitations attached to it.

Exemption

The law[2] allows the Department to grant exemptions from economic certificate requirements when it finds that granting such exemptions is "consistent with the public interest." Exemptions may be granted to individual air carriers or to groups or classes of air carriers. Exemption is also issued to companies that already hold certificate authority but who wish to provide operations outside the authority granted by the certificate. For example, Burning Stump Airlines holds a domestic certificate for scheduled passenger operations, but no foreign air transportation. It may be granted an exemption to provide foreign cargo service.

Operations with small aircraft

Carriers proposing to operate only small aircraft[3] are exempt from certificate requirements. They obtain economic authority as air taxi operators or commuter air carriers. Air taxi operator and commuter air carrier are registrations rather than certificates of public convenience and necessity. They operate under the provisions of 14 CFR Part 298.[4]

[2]49 USC 40109. Regulations governing exemptions are addressed in 14 CFR Part 298.

[3]Aircraft with 60 or fewer seats or a payload capacity of 18,000 pounds or less.

[4]In 1995, the FAA and the Assistant Secretary for Aviation and International Affairs agreed to transfer Part 298 Exemption Authority to the FAA. The function is carried out by the FAA Flight Standards Service.

The economic certification process

Applicants for a certificate must be U.S. citizens. A three-part test is used to determine whether a company is "fit, willing and able."

- First, managerial competence of the applicant's key personnel is examined to determine whether they have sufficient business and aviation experience to operate an airline, and whether the management team, as a whole, possess the background and experience necessary for the specific kind of operations proposed.
- Second, the applicant's operating and financial plans are submitted in order to judge whether the applicant has a reasonable understanding of the costs of starting its operations and has a specific and realistic plan for raising the necessary capital. The applicant must submit third-party verification that it has acquired the necessary capital.
- Third, the applicant's compliance record is examined to see whether it and its key personnel have a history of safety violations or consumer fraud activities that would pose a risk to the traveling public.

The submitted data[5] are docketed.[6] Interested persons may file answers to the application either supporting or opposing it.

The Department's Air Carrier Fitness Division reviews the application. If there are no issues inhibiting approval of the application, a "show cause" order is published in the *Federal Register* allowing interested persons to file comments and "show cause" why the Department should or should not adopt its proposed fitness findings and award of authority. If no objections are filed a final order is issued awarding the applicant a certificate. The certificate may have limitations attached to it, such as restrictions on the type of service or the number and type of aircraft that may be used.

If an application is to provide service to a foreign point, the final decision to grant or deny authority is with the President of the United States.

[5]14 CFR Part 204 defines the minimum evidence to be presented to the Department upon which it evaluates the applicant. However, the Department may require additional data.

[6]A docket in this sense is a file of the relevant documents and evidence associated with the application. It is available to the public. Sensitive proprietary and competitive information submitted may be withheld from the public.

Operating authority

The Federal Aviation Administrator issues operating certificates and Operations Specifications. In order to conduct air carrier operations under 14 CFR Part 121, an applicant must be a citizen of the United States of America.

There are five phases in the certification process. Each phase is described in sufficient detail to provide a general understanding of the entire certification process. The five phases are:

1. Preapplication
2. Formal application
3. Document compliance
4. Demonstration and inspection
5. Certification

The operating certification process

Certificates are issued to United States citizens who meet the requirements of 14 CFR Part 119 and who have appropriate economic certificates issued by the Department of Transportation.

There are four basic types of air operator certificates. They define the kinds of operations and the rules under which the certificate holder will operate. These are *Domestic, Flag, Supplemental, Commuter,* and *On-Demand Commercial.* The kind of a certificate an airline may hold is determined by applicability tests defined in the regulations. The tests center about the type of equipment to be flown, passenger seating capacity, flight frequency, and geographic areas of operation. It is possible to hold more than one kind of authority.

The regulations dictate that an application for a certificate is made[7] "In a form and manner prescribed by the Administrator; and…containing any information the Administrator requires the applicant to submit."[8] The form is highly structured and ritualized. It consists of five phases and three gates (Figure 3-1) in the certification process.

[7]Each applicant must submit the application to the Administrator at least 90 days before the date of intended operation. 14 CFR Part 119.35.

[8]This is described in detail in AC 120-49, Certification of Air Carriers.

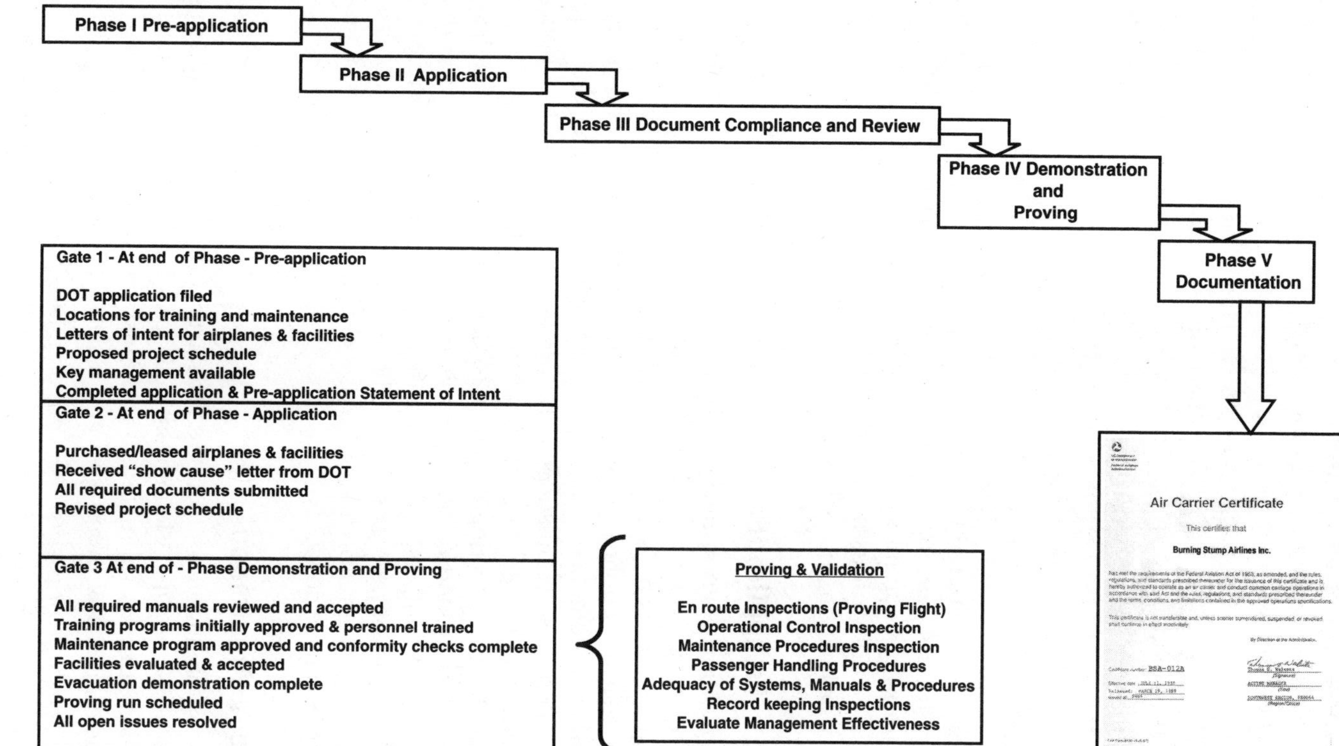

Figure 3-1 *Operating Certificate application process.*

- *Phase I, preapplication, and Phase II, formal application.* These two phases constitute the initiation of the program and its formal definition including preliminary task definition and schedules.
- *Phase III, document compliance.* This is the detailed stage of the project. Plans are solidified and expanded, detail tasks are identified, required materials are prepared, facilities are finalized, and compliance statements prepared and submitted to the FAA for review and acceptance.
- *Phase IV, demonstration and inspection.*
- *Phase V, certification.*

Three gates require the completion of specific items before continuation of the process into another phase or specific event. The gate system ensures continuity between phases and avoids transitioning into and out of various phases by all the parties to the process. It assures coordination and continuity.

Phase I—Preapplication

Preapplication is an applicant's initial contact with the FAA. During these meetings, only basic high-level information and general certification requirements are discussed. A high-level briefing is given to ensure that the applicant fully understands the certification process and the requirements. Preliminary discussions regarding the applicant's intentions are included. The applicant is given a Preapplication Statement of Intent to complete and return to the FAA. After this, a preapplication meeting(s) is held to define the project. Practical issues such as ground rules, a preliminary plan for the project, formal and informal contacts, anticipated issues, and procedures are established. Details of prerequisites for the specific formal application are discussed. The preapplication phase is completed with the first of the three gate reviews.

Phase II—Application

After completing Gate 1, to the satisfaction of all parties, the application phase begins. A letter that includes a request to be certificated as an air carrier is submitted. It includes attachments that define the significant items of the project and a schedule of events.

- *Schedule of events.* The schedule of events lists items, activities, programs, and aircraft and/or facility acquisitions that must be accomplished or made ready for FAA inspection before certification.

- *Company general manuals.* These manuals or sections of manuals contain information about the applicant's general policies, duties, and responsibilities of personnel, operational control policy, and procedures. These are commonly referred to as the General Operations Manual and the General Maintenance Manual. Many of the items required for compliance will be contained in the manuals.
- *Initial training curriculum.* The company's initial training curriculum (completed to the extent possible) must be attached to the formal application letter.
- *Management qualification résumés.* 14 CFR Part 121 establishes basic management positions and minimum qualifications for air carrier applicants. This attachment must contain résumés of the qualifications, certificates (including certificate numbers), ratings, and aviation experience for each of the required individuals
- *Documents of purchase, leases, contracts, and/or letters of intent.* These provide evidence that aircraft, facilities, and services have been acquired. If formal arrangements are not completed letters of intent suffice until formal contracts or agreements become available.
- *Initial compliance statement.* This is a listing of all 14 CFR sections applicable to the proposed operation. Compliance statements show how the applicant conforms to the regulations. They are voluminous. At this stage, however, the statement is essentially a plan. The brief description or reference describes the method of compliance for each regulation listed.

At the formal application meeting expectations and project details are discussed. The plan and schedule details are refined. Specific regulatory and special requirements are confirmed. Conformity issues are opened including such items as exemption or deviation from regulatory requirements. Housekeeping, methodologies and ground rules, and formal and informal contacts for the project are secured.

It is unconditionally essential when establishing the project that only those items specifically required by regulation or law are included in the schedule plan and compliance statement. It is not sufficient to identify the regulatory compliance items. Laborious definition of what shall constitute compliance must be agreed to between the parties as well. The project must be unequivocally bounded. This avoids later misunderstandings between the parties and the introduction of de facto regulation beyond the authority of the regulator.

Much of the work associated with establishing an air carrier operation is not related to the certifying 14 CFR 119. Numerous other agencies will have some jurisdiction over the conduct of operations. Consequently, it is possible that more than one plan and schedule are developed to meet all of these conditions. The regulatory requirements of these agencies and attendant compliance must be kept segregated from each other. The same may be said for the development of business processes that are not directly connected to a regulation.

Before proceeding to Phase III the FAA will hold the second gate review with the applicant. Particular items addressed in the review follow: the economic authority application must have proceeded to the "show cause" stage, airplanes and facilities must have been purchased/leased, all required documents must have been submitted, and a detailed schedule and plan must have been completed.

Phase III—Document Compliance and Review[9]

This phase contains the substance of the items presented in the plan and schedule of events. The compliance statement is a detailed item-by-item statement concerning how the applicant should conform to each specific item of the applicable regulations. It is an enormous body of material consisting of voluminous manuals, statements, exhibits, and processes. Some highlights included in the statement include:

> Acceptance of information, which does not require formal approval, will be indicated by letter or by the lack of any FAA objections to the information.
>
> Typical compliance topics are shown in Figure 3-2. This is by no means a complete list of items addressed in the compliance statement.

[9]Although the document compliance and the demonstration and inspection phases are discussed separately, these phases overlap. They are accomplished simultaneously in actual practice.

Typical Compliance Topics

Common Items

General policies, administrative processes, and procedures for the conduct of flight operations and maintenance including reference to appropriate Federal Aviation Regulations.

A complete list of the aircraft to be operated must be submitted and published in appropriate manuals and the Operations Specifications. This includes identity by aircraft serial number. Airplanes may only be listed in the Operations Specifications of a single carrier. This assures among other things the clear delineation of what approved maintenance program and under whose control the airplanes will exist.

Training programs including curricula for flight crew, flight attendant dispatch, and maintenance personnel.

Accident procedures addressing notification and response procedures, preservation of wreckage, records, etc.

Duties and responsibilities of crew members, appropriate members of the ground organization, and management personnel.

14 CFR 121 is specific about the management positions that must be established within the airline. They define the positions and outline the minimum qualifications the person filling the position must meet. As part of the compliance, these individuals must be identified and a résumé submitted as part of the certification process. These résumés normally include prior associations and regulatory history.

Appropriate information from the operations specifications must be made available to employees. This usually takes the form of reproducing sections of the specifications in various manuals.

Facilities and services.

Typical Compliance Topics

Manuals. The manuals may be divided into separate documents appropriate to specific audiences. The most commonly referred manuals in the regulations are the General Operations and General Maintenance Manuals. These manuals define the general administrative operating practices and processes used to comply with the regulations and run flight and maintenance programs.

Weight and balance control and loading. This includes a weighing program, records systems and methods for determining airplane weight and center of gravity by dispatch or flight crews

Minimum Equipment List (MEL). This includes the MEL itself and the administrative practices associated with its use including crew notifications and control of time limited items. See Chapter 7.

Figure 3-2 *Typical compliance topics.*

Flight Operations

Flight dispatch/flight following and operational control procedures.

Airplane operating instructions including system operation and flight maneuvers. This is traditionally known as the airplane operations manual. The manufacturer's operations manual is most often used as a starting point. However there are flight operations procedures associated with the airplane which must be developed by the airline. An example is the emergency evacuation plan developed by the airline.

Airport and route information including en route, flight, navigation, and communication procedures.

Airplane performance information covering takeoff, en route, and landing weight limitations. This material is based upon the performance information contained in the airplane Approved Flight Manual. Today this material is computer generated and is referred to as the airport analysis document.

Procedures for familiarizing passengers with the use of emergency equipment during flight (passenger briefing and safety cards).

Emergency equipment and procedures. This includes procedures for the use of the equipment and evacuation and ditching procedures

The method of designating succession of command of flight crew members.

Procedures for operating in periods of ice, hail, thunderstorms, turbulence, or any potentially hazardous meteorological conditions.

Pilot and dispatcher route and airport qualification procedures.

Maintenance Operations

Maintenance reliability program.

Continuous analysis and surveillance system.

Instructions and procedures for performing maintenance, preventive maintenance, and servicing. This is normally contained in the manufacturer's maintenance documents. It is commonly referred to as the Airplane Maintenance Manual. It includes such topics as removal and installation, cleaning, inspection, repair, overhaul, and test. Maintenance documents for a transport category airplane are typically a large group of manuals. The manufacturer and component vendors prepare them.

Time limitations, or standards for determining time limitations, for overhauls, inspections, and checks of airframes, engines, propellers, appliances, and emergency equipment. These items define the basic inspection program. Their origin is the Maintenance Review Board Report for the airplane type design. They will be contained in Part D of the Operations Specifications for each airplane type operated.

The method of performing routine and nonroutine maintenance (other than required inspections), preventive maintenance, and alterations. Much of this material is contained in the airplane maintenance publications. However administrative procedures and records management must also be included.

Figure 3-2 *(Continued)*

Identification and handling of Required Inspection Items (RII). This includes designation of personnel authorized to perform RII; acceptance/rejection criteria for the inspections; procedures to ensure that the inspections are accomplished. Required Inspections are normally associated with items that if improperly completed can result in a serious compromise to safety.

Shift change procedures are associated with practices to assure that continuity exists between outgoing and incoming shift personnel (particularly inspection).

A quality assurance program including the authority of inspectors; protection from inspectors being overridden and reinspection and buyback procedures. QA programs also include methods and procedures for continuing surveillance and analysis of the maintenance program. Airworthiness inspections, including instructions covering procedures, standards, responsibilities, and authority of inspection personnel.

Typical Compliance Topics

A records system including a record of AD note accomplishment that provides for preservation and retrieval of information in a manner acceptable to the Administrator and that provides

(1) A description (or reference to data acceptable to the Administrator) of the work performed

(2) The name of the person performing the work if the work is performed by a person outside the organization of the certificate holder

(3) The name or other positive identification of the individual approving the work

Procedures for refueling aircraft, eliminating fuel contamination, protection from fire (including electrostatic protection), and supervising and protecting passengers during refueling.

Airplane configuration control.

Tool calibration program.

A list of agents with whom it has arranged for the performance of any of its required inspections, other maintenance, preventive maintenance, or alterations, including a general description of that work. This defines third-party vendors with whom agreements have been entered into.

Methods and procedures for maintaining the aircraft weight and center of gravity including appropriate administrative practices and records management.

Hazardous Materials Program

Procedures and information to assist personnel to identify packages marked or labeled as containing hazardous materials and if these materials are to be carried, stored, or handled, procedures and instructions relating to the carriage, storage, or handling of hazardous materials, including the following:

Procedures for determining the proper shipper certification required by 49 CFR Subchapter C, proper packaging, marking, labeling, shipping documents, compatibility of materials, and instructions on loading, storage, and handling.

Figure 3-2 *(Continued)*

Notification procedures for reporting hazardous material incidents as required by 49 CFR Subchapter C.

Instructions and procedures for the notification of the pilot in command when there are hazardous materials aboard, as required by 49 CFR Subchapter C.

Other information or instructions relating to safety.

Drug Testing Program

49 CFR Part 40 and appendices in 14 CFR 65, 121, and 135 mandate drug testing of persons who perform a safety-sensitive function. This includes flight crew members, flight attendants, and mechanics.

Alcohol Misuse Prevention Program

14 CFR 121 requires a program that is designed to help prevent accidents and injuries resulting from the misuse of alcohol by employees who perform safety-sensitive functions in aviation. This includes appropriate training and guidance for employees and supervisors. It includes alcohol testing procedures.

Safety Program

See AC 120-66, Aviation Safety Action Programs (ASAP).

Security Program

14 CFR 108 mandates that air carriers implement a security program. This includes training of appropriate personnel, screening of passengers, carriage of law enforcement personnel, carriage of prisoners, carriage of weapons and the security of airplanes and facilitates. Of particular interest is the process for the security of facilities and airplanes.

Figure 3-2 *(Continued)*

Phase IV—Demonstration and Proving

Phase IV consists of demonstrating, before revenue operations, the ability to safely operate using the proposed flight and maintenance operations programs. Demonstrations include the actual conduct of operations. The FAA evaluates the effectiveness of management and the processes established. Many of the items are evaluated during the proving flight. Examples of the types of items, equipment, facilities, and activities validated during this phase include:

- Training programs including instruction, facilities, simulators, aircraft, and instructor personnel
- Air personnel certificate verification
- Station facilities, procedures, and personnel

- Record-keeping procedures including maintenance, training, crew flight and duty times, flight papers, aircraft records, AD (Airworthiness Directive) note accomplishment, required inspection items, etc.
- Dispatch/flight following capabilities
- Maintenance, inspection programs and facilities including personnel, tooling, technical information, spares capability, fueling, ramp operations, maintenance procedures
- An airplane[10] conformity inspection. The airplane must be in U.S. registry, must be airworthy and include the equipment required by 14 CFR Part 121, must have completed aircraft maintenance records, etc. Examples of required equipment include communications and navigation capability.
- Minimum Equipment List (MEL) practices and procedures.
- Use of the Configuration Deviation List (CDL).
- Weight and balance.
- Publications control, distribution, and revision practices.

Proving, tests, and demonstrations

The FAA requires a demonstration of an operator's capability. This applies to both new operators seeking certification and to certificated operators changing their operating environment, equipment, and so forth. The regulatory and procedural aspects of these are voluminous and detailed. These tests should not be confused with aircraft certification tests, which are done by the aircraft manufacturer to demonstrate the airworthiness of the aircraft.

Validation tests

Validation tests are conducted to demonstrate capability to operate over specific routes while using specific navigational equipment, or to operate within specified limitations in critical areas. Although proving flights and validation tests satisfy different requirements, it is

[10]Subpart H of 14 CFR Part 121 describes aircraft requirements. It is possible to operate a leased or charted airplane that is in foreign registry, with restrictions. The airplane must be maintained under the airworthiness standards and maintenance program approved by the country of registry. Specific limitations will be listed in the operator's Operations Specifications. The first example of this occurred in the early 1980s when 737 airplanes were leased back and forth between Air Florida (a U.S. carrier) and Air Europe (a U.K. carrier). The reasons both carriers exchanged airplanes was sound. When Air Florida needed additional lift Air Europe was in a slack revenue period and vice versa. Unfortunately neither operator is currently in business.

common to conduct both tests simultaneously. Validation tests may involve flight demonstrations.

Examples of validation tests that affect maintenance operations include the initiation of Category II and III operations, ETOPS, RNAV operations, autoland operations, and log range navigation operations, and special performance such as operation from high-altitude airports.

Typical maintenance activities associated with tests are adequacy of MEL coverage, appropriate equipment certification, maintenance program coverage, records coverage, personnel qualification, and coordination of activity between flight operations and maintenance.

Evacuation demonstrations

Passenger emergency evacuation demonstrations are conducted to validate that, using the operator's procedures and flight and cabin crews, the full capacity of the airplane will be evacuated in 90 seconds or less.[11]

Full-scale demonstrations using an airplane and a full complement of representative "passengers" were the norm until recently. Every operator was required to conduct the demonstration any time a new airplane type was placed into operation or if the seating capacity of previously demonstrated configurations changed more than 5 percent. Full-scale demonstrations are extremely costly and carry the risk that the participants may be injured.

Contemporary requirements are such that usually only a new operator is required to conduct a full-scale demonstration. If another 14 CFR Part 121 operator or a manufacturer as part of the airplane certification has previously completed a full-scale demonstration, only a partial demonstration is required of operators. The exact details and depth of the demonstration are negotiated with the FAA.

Ditching demonstrations

If extended overwater operations are to be conducted the operator must do a ditching demonstration. This usually consists of a simulated ditching using a typical line crew. The purpose is to demonstrate the validity of the ditching procedures, crew training, and reliability of the ditching equipment. It usually consists of removing

[11] 14 CFR Part 121.291 and Appendix D. Also see 14 CFR Part 25.803.

each lift raft from its stowage, one raft is inflated and launched (or a slide raft is inflated) and the crew members assigned to that raft demonstrate the use of the emergency equipment.

Proving flights

Proving flights[12] are conducted to show that an operator is capable of operating a specific aircraft type. The flights consist of a full-scale simulation of revenue operations. Proving tests consist of a demonstration of the applicant's ability to operate and maintain an aircraft new to the fleet or the applicant's ability to conduct a particular kind of operation, such as domestic, flag, or commuter. The applicant is required to operate and maintain the aircraft to the same standards required of a certificate holder that is fully certificated and that holds the necessary authorizations. If the airplane to be used is already in air carrier service, the number of proving hours is 50. The number of night flying hours is negotiated at the time the proving flight plan is submitted to the FAA. Flights must be conducted over a representative portion of the operator's routes and into en route airports. Revenue passengers may not be carried during proving flights. However, cargo may be carried.[13]

If an airplane has not previously been used in air carrier operations, the number of flight hours is 100.[14] These types of proving are not just testing the ability of the operator but also the ability of the airplane. They are called Initial Proving Tests. For example, the first operator of the Boeing 747 was Pan American. They were required to conduct 100 hours of proving rather than 50.

Maintenance program approval

Details of the elements of the maintenance program are discussed in Chapter 12.

Phase V—Documentation

After the previous phases and gates have been met, the only remaining task is to issue the Operating Certificate and Operations Specifications. Before this is done the operator must supply a copy of its economic authority from the Office of the Secretary of Transportation.

[12] 14 CFR Part 121.163.

[13] Applicants seeking FAA certification that do not have appropriate DOT economic authority are not permitted to carry revenue cargo.

[14] Ten of the hours must be at night.

The Air Carrier Certificate, the Operations Specifications, and the economic authority enable the certificate holder to begin revenue operations. Economic authority is not effective until after FAA issues both the Air Carrier Certificate and the Operations Specifications.

Operations Specifications

Until the early 1950s, the Operating Certificate contained specific limitations on the conduct of operations. These limitations are now split for easier administration and control. Certificates issued under 14 CFR 119 stipulate that operation

- Must be conducted in accordance with the Operations Specifications
- May not be conducted without Operations Specification

In practice the Operations Specifications are considered "what makes you an airline." They define the way the airline will conduct its flight and maintenance operations, defining specific requirements and limitations. Operations Specifications are not considered part of the Operating Certificate[15] but you may not conduct operations without them.

As an air carrier's operation changes, the Operations Specifications will be amended accordingly. The process for amending Operations Specifications is similar to the certification process.

Need for Operations Specifications

It is impractical to develop regulations that fit every conceivable environment. More practical is to tailor the intent of the regulations to a specific operator. This is the concept of Operations Specifications. Operations Specifications customize the operating rules to the operating environment, routes and areas of operations, operator experience and capability, maintenance facilities, and organizational structure. Operations Specifications carry the full weight of regulation.

Contents of Operations Specifications

14 CFR 119.7 and 119.49 define the contents of Operations Specifications. A summary is shown in Figure 3-3.

[15]Except for Operations Specifications paragraphs identifying authorized kinds of operations.

The specific location of the certificate holder's principal base of operations and the name and mailing address of the certificate holder's agent for service

Other business names under which the certificate holder may operate

Reference to the economic authority issued by the Department of Transportation

Type of aircraft, registration markings, and serial numbers of each aircraft authorized for use

Each regular and alternate airport to be used in scheduled operations, and, except for commuter operations, each provisional and refueling airport

Kinds of operations authorized

Authorization and limitations for routes and areas of operations

Airport limitations

Time limitations, or standards for determining time limitations, for overhauling, inspecting, and checking airframes, engines, propellers, rotors, appliances, and emergency equipment

Authorization for the method of controlling weight and balance of aircraft

Interline equipment interchange requirements

Aircraft wet lease information

Any authorized deviation and exemption granted

Any other item the Administrator requires

Figure 3-3 *Operations Specifications contents.*

Format of Operations Specifications

Operations Specifications are divided into six parts (Figure 3-4). Each addresses specific topics. Individual parts are composed of a series of standard paragraphs.

An automated Operations Specifications program has been implemented. Safety inspectors enter appropriate information about the operator into a computer database. The program in turn produces a customized specification that includes only those authorizations, limitations, standards, and procedures applicable to an individual certificate holder. Special authorizations and limitations may be approved for a particular certificate holder. The program thus allows nonstandard paragraphs to be added that address unique conditions.

Each page of an Operations Specification has signature space for the responsible FAA inspector and for the operators. These signatures indicate acceptance of the contents of the page by both parties.

Part	Title	Typical Subjects Addressed
A	General	• Issuance and applicability • Definitions and abbreviations • Airplane authorizations including make, model, series, serial number, passenger seating capacity, and required number of flight attendants • Exemptions and deviations • Management personnel • Other designated persons including agent for service and person(s) authorized to apply for and accept ops specs • Operational Control • Airport aeronautical data • Aeronautical weather data • Aircraft wet leases • Interchange agreements, special authorizations, and limitations
B	En route Authorizations, Limitations, and Procedures Operations	• Authorized areas of en route operation, limitations, and procedures • Flight rules limitations and provisions • Conduct of extended range operations with two engine airplanes • Conduct of IFR operations in uncontrolled airspace • Redispatch and rerelease • Class 1 navigation using area navigation systems in U.S. positive control area using area or long-range navigation systems • Class 2 navigation using long-range navigation systems or a flight navigator • Operations in areas of magnetic unreliability • Operations within the North Atlantic Minimum Navigation Performance Specifications (MNPS) • North Atlantic operation with two-engine airplanes • Extended range operations with two-engine airplanes under FAR 121 • Special fuel reserves in international operations • Planned inflight redispatch or rerelease en route

Figure 3-4 *Operations Specifications parts.*

Part	Title	Typical Subjects Addressed
C	Airplane Terminal Instrument Procedures and Airport Authorizations and Limitations pertaining to airplanes only	• Terminal instrument procedures • Basic instrument approach procedures • IFR landing minimums other than Categories II and III • Special limitations • IFR takeoff minimums • CAT II and III operations • Approach operations using area navigation • Instrument approach procedures using area navigation • Powerback operations with airplanes • Takeoff operations in a tailwind • Airport authorizations
D	Aircraft Maintenance	• General • Check, inspection, and overhaul time limits • Use of an MEL • Reliability program authorization • Short-term escalation authorization • Maintenance contractual agreements • Leased aircraft maintenance program • Parts pool agreements • Parts borrowing • Special flight permit with continuing authorizations • List of aircraft operated • Approved aircraft inspection program • Additional maintenance requirements
E	Weight and Balance	• Weighing program • Methods for determining weight and center of gravity
H	Helicopter Terminal Instrument Procedures and Airport Authorizations and Limitations	

Figure 3-4 (*Continued*)

Issues affecting certification and operation

Code sharing

Code sharing is a marketing arrangement that permits an air carrier to sell service under its name and airline designator code when the service is provided in whole or in part by another air carrier; for example, a foreign air carrier. The converse is also true. Code sharing is a marketing arrangement requiring Department of Transportation (DOT) approval.

Three types of arrangements may be authorized:

- *Type 1.* After the foreign carrier operates its own service between homeland and a point in the United States, the U.S. partner, under code sharing, transports the foreign carrier's passengers within the United States.
- *Type 2.* After the U.S. carrier operates its own service between the United States and a foreign point, the foreign partner, under code sharing, transports the U.S. carrier's passengers between foreign points.
- *Type 3.* U.S. carrier's passengers are transported from the United States to a foreign point in a U.S. or foreign carrier's aircraft.

Recently the DOT has been withholding its approval until receiving FAA confirmation that a foreign civil aviation authority (CAA) is providing oversight of the foreign air carrier in accordance with minimum ICAO standards. The FAA bases its confirmation on an assessment of the CAA's capabilities.

Alliances

An alliance may involve a merging of resources, operations, or financial interests between a U.S. and a foreign-certificated air carrier or repair station. The Star Alliance between United Airlines and several foreign carriers is an example. It could involve the sharing of parts or the use of mechanics, pilots, and flight attendants. Such arrangements are more complex than the simple marketing relationship established under a code-sharing agreement. They involve relationships that affect maintenance operations.

- Crew interchanges—pilots and flight attendants
- Aircraft interchanges
- Common aircraft maintenance programs
- Common aircraft reliability programs
- Aircraft parts interchanges
- Exchange of maintenance personnel and facilities
- Audit and quality control
- Training programs

3

Airworthiness

4

Airworthiness—General

Legislative basis

The legislative authority to establish and control safety has been secure in United States law ever since the adoption of the Air Commerce Act of 1926. Many of the original reasons for the passage of the 1926 act surrounded a demand from within the aviation industry for regulation, including the establishment of standards for equipment and air personnel. Many of the equipment standards established in 1926 were adopted from regulations used by the United States Post Office. These formed the core of modern airworthiness regulations.

Presently the FAA Administrator sets the standards for airworthiness. Contemporary legislation states in part that "The Administrator…shall promote safe flight…by prescribing…standards required in the interest of safety for appliances and for the design, material, construction, quality of work, and performance of aircraft, aircraft engines, and propellers."[1]

Airworthiness defined

At best, airworthiness is a sensitive, contentious word. The very word unfortunately generates emotional outbursts, heartburn, occasional unnecessary certificate action, and confusion. There seem to be as many definitions as there are people. Some definitions seem to have their origins in illusion. "It's what makes a flying machine worthy of the air." "It's airworthy when it looks right." "Its airworthy when it contains everything the manufacturer's drawing say should be installed."

[1]49 USC 44701.

"Airworthiness is the ability to make the trip safely." Nonsense! It is none of these. There is only one definition for airworthiness.

What makes a design airworthy? Has an Airworthiness Directive (AD) note been conformed? Has a required inspection been performed? If the performance of an engine has deteriorated beyond its new condition, is it airworthy? Substantial confusion, myth, and emotion surround the definition of airworthiness. This flaws the application and management of the notion.

Airworthiness is not an aerial vehicle that possesses certain properties of aerodynamic efficiency. It has nothing to do with favorable thrust-specific fuel consumption, low weight, etc. It is not a design that "looks good." Airworthiness is a property of aeronautical products. But what defines it?

It is clearly defined by the court and the FAA. In 1991, the Administrator defined it as follows:

> *The Administrator has held that an aircraft is airworthy when: (1) it conforms to its type design or supplemental type design or to any applicable airworthiness directives, and (2) is in a condition for safe operation.*[2]

Airworthiness is not a state of being. It is not conditional. An airplane is either airworthy and its operation is safe or it is not. You will avoid considerable emotional and professional turmoil if you stick rigidly close to the Administrator's definition. No other definition applies. Don't be influenced by people who start their arguments by saying. "It is my/our understanding/policy...." Do not allow personal feeling and interpretations or local tribal wisdom to enter discussions of airworthiness. Doing so draws one into a quagmire of confusion and emotion from which it is impossible to extricate oneself. A pragmatic application of the definition avoids the trap. Above all else, don't be moralistic or philosophical about airworthiness. Airworthiness is neither moral nor philosophical. It is a regulatory issue!

The maintenance community perceives airworthiness as being solely airplane related. In fact, there are numerous citations in the literature and FAR that promotes this view. But note well, airworthiness is de-

[2]In the matter of Watts Agricultural Aviation, FAA Order No. 91-8, at 17(April 11, 1988) [citing Section 603(c) of The Federal Aviation Act of 1958, as amended, 49 U.S.C. App.1423(c)], appeal docketed, No. 91-70365 (9th Cir. 1991).

fined by two conditions in the Administrator's definition. Let's examine both.

Conformity

The Basis of Certification is established before the start of certification of an airplane design. It defines the applicable FAR, special conditions, and exemptions that the manufacturer must use to demonstrate the airworthiness of an airplane design. Showing, during certification, what devices, system(s), features, or performance in the design fulfill these requirements demonstrates conformity (airworthiness) in the design. The basis of certification is listed in the airplane, power plant, or propeller Type Certification Data Sheet.[3]

We tend to equate airworthiness with safety. This serves us well. But it is not completely valid. Several airworthiness rules address environmental concerns rather than safety issues. For example, if an airplane does not meet the noise rules it is not airworthy. It does not conform. Is it capable of safe flight? Yes, it is just too noisy.

Conformity is at the heart of airplane airworthiness. Assuring continued conformity is the duty of the maintenance community.

Safe operation

Safe operation relates to the conduct of a given flight operation. It is the responsibility of the pilot in command[4] (or, in some cases, a joint responsibility with a certificated dispatcher) to determine safe operation.

Safe operation is determined by the following factors:

- *The airplane is airworthy.* The airplane must be in full conformity to the applicable airworthiness rules. Certificated maintenance personnel make this determination and convey it to the pilot in command.
- *The airplane is capable of safe flight.* The airplane may be in full conformity and still not be capable of safe flight. A simple example is the case of an airplane coated with ice and snow.

[3]Technical Standard Orders (TSO) define airworthiness for appliances certificated under the TSO system. Examples include seat belts, brakes, wheels, and tires.

[4]14 CFR 91.7 states in part "The pilot in command of a civil aircraft is responsible for determining whether that aircraft is in condition for safe flight."

- *Installed equipment.* The airplane must be properly equipped for the proposed operation. Examples include:
  - Operative anti-icing equipment into areas of known icing.
  - Equipment for extended twin operation. This can include a verification that the airplane meets ETOPS requirements.
  - Navigation equipment appropriate to the route to be flown as well as the planned type of landing approach. This can include verification that the airplane is capable of operation at lower minima (CAT II or III, for example).
  - Communication equipment appropriate to the planned route of flight. For example, HF radios for extended overwater operations.
- *Other factors.* Those additional items required for safe flight include, but are not limited to, the following examples:
  - Takeoff, en route, and landing airplane performance
  - Weight and center of gravity within acceptable limits
  - Takeoff, en route, and landing meteorological conditions including instrument approach minima within limits
  - A properly executed flight plan including fuel requirements
  - Knowledge of the status of navigation aids to be used

Economic features

It is quite common for designs to incorporate features that promote ease-of-flight operations, fuel efficiency, maintenance, passenger comfort, appearance, and the like. Examples include built-in diagnostic systems (maintenance computers), interior furnishings, cockpit clipboards, coat closets, flight management computers, and so forth. These are economic features of the design.

Although the incorporation of these features into a design may carry airworthiness implications, their requirement for flight operations is not required. For example window shades or curtains in the main cabin of a transport aircraft are not required by the airworthiness regulations. However, the materials used for these must conform to the flammability requirements of the airworthiness regulations.

Applicability of airworthiness regulations

International Civil Aviation Organization (ICAO) Annex 8, Airworthiness, outlines the basic recommendations of the Chicago Convention regarding the subject. Signatories to the convention recognize the airworthiness of the country of origin. However, under the convention, a receiving country may impose special airworthiness conditions upon the airplane. In the past it was common practice to impose a device called a stick nudger. The device is required under U.K. airworthiness rules. Thus to import a 707 into U.K. registry the airplane has to have the stick nudger installed.

Airworthiness rests with the country of original certification. Continuing airworthiness, however, is the responsibility of the country of registry. This leads to interesting dilemmas during the life of an airplane. Consider, for example, a 747 airplane certified and manufactured in the United States, which is purchased by British Airways. The airplane is placed into U.K. registry. The U.K. continuing airworthiness rules apply to the airplane. Furthermore, consider now that installing a new passenger cabin interior modifies the airplane. The new interior is certified under U.K. airworthiness rules.

Now let's confuse the topic. Let's suppose the airplane is subsequently sold to Lufthansa and placed into German registry. The airworthiness of the airplane is now defined by German rules. But some features certified on the airplane while it was in U.K. registry do not meet the airworthiness rules of Germany. Now what? The airplane would be altered to bring it into conformity with German rules.

Responsibility for airworthiness

The regulation is quite specific: "The owner or operator of an aircraft is primarily responsible for maintaining that aircraft in an airworthy condition, including compliance with Part 39 of this chapter."[5]

Further air carrier certificate holders are responsible for:

- *The airworthiness of its aircraft, including airframes, aircraft engines, propellers, appliances, and parts thereof; and*

[5]14 CFR 91.403.

- *The performance of the maintenance, preventive maintenance, and alteration of its aircraft, including airframes, aircraft engines, propellers, appliances, emergency equipment, and parts thereof, in accordance with its manual and the regulations of this chapter.*[6]

Section 44702(b) of Title 49 specifies, in part, that when prescribing standards and regulations and when issuing certificates, the FAA shall give full consideration "to the duty resting upon air carriers to perform their services with the highest possible degree of safety in the public interest."

It is possible to arrange with another person for the performance of any maintenance, preventive maintenance, or alterations. However, this does not relieve the responsible agents of this responsibility regardless of the type of operations.

Airworthiness regulations

14 CFR 21 establishes the standards applicable to all categories of airworthiness. It lays out the basic rules and administrative considerations. Our principle concerns are with air carrier operations and thus we are predominantly concerned with 14 CFR 21, 25, 33, 34, and 39. The regulations are being harmonized with the corresponding JAR issued by the Joint Airworthiness Authorities. A summary of airworthiness regulations is shown in Figure 4-1.

Airworthiness design standards

14 CFR 25.1309 requires that:

> *(a) The equipment, systems, and installations whose functioning is required by this subchapter, must be designed to ensure that they perform their intended functions under any foreseeable operating condition.*
>
> *(b) The airplane systems and associated components, considered separately and in relation to other systems, must be designed so that -*
>
> > *(1) The occurrence of any failure condition which would prevent the continued safe flight and landing of the airplane is extremely improbable, and*

[6]14 CFR 121.363.

United States Federal Aviation Regulations—Airworthiness			
Administrative	**Airframe**	**Power Plant**	**Environmental**
FAR Part 21 Certification Procedures for Exhaust Products and Parts	**FAR Part 23** Airplane Airworthiness Standards: Normal, Utility, Acrobatic, and Commuter Category Aircraft	**FAR Part 33** Airworthiness Standards: Aircraft Engines	**FAR Part 34** Airworthiness Standards: Fuel Venting and Emission Requirements for Turbine Powered Airplanes
	FAR Part 25 Airplane Airworthiness Standards: Transport Airplanes Category	**FAR Part 35** Airworthiness Standards: Propellers	**FAR Part 36** Noise Standards: Aircraft Type and Airworthiness Certification
	FAR Part 27 Airworthiness Standards: Normal Category Rotorcraft		
	FAR Part 29 Airworthiness Standards: Transport Category Rotorcraft		
	FAR Part 31 Airworthiness Standards: Manned Free Balloons		
FAR Part 39 Airworthiness Directives			

Figure 4-1 *List of all airworthiness FAR.*

(2) The occurrence of any other failure conditions which would reduce the capability of the airplane or the ability of the crew to cope with adverse operating conditions is improbable. Airworthiness design standards are based upon design strategy known as fail-safe. It considers the effect of failures and combinations of

failures when defining a design. Fail-safe strategies insure that major failures are improbable and that catastrophic failures are extremely improbable....

Essentially this means that no single failure of structures, components, or systems may imperil the airplane or its occupants.

The fundamental premises are embodied in a strategy called fail-safe design.

- Single failures of any component or system during any one flight is assumed, regardless of its probability. These single failures cannot prevent continued safe flight and landing, or significantly reduce the capability of the airplane or the crew to cope with the failure.
- Later failures during the same flight, whether detected or latent, and combinations thereof, are assumed, unless their joint probability with the first failure is shown to be extremely improbable.

Fail-safe design uses a combination of design methods.

1. Design integrity and quality, including life-limits, to ensure intended function and prevent failures.
2. Redundancy, fault tolerance, or backup systems to enable continued function after any single (or other defined number of) failure(s), for example, two or more engines, hydraulic systems, flight control systems.
3. Isolation of systems, components, and elements so that the failure of one does not cause the failure of another. Isolation is also termed independence or brick walling.
4. Proven reliability so that multiple, independent failures are unlikely to occur during the same flight.
5. Failure warning or indication to assure failure detection.
6. Flight crew procedures for use after failure that enable continued safe flight and landing.
7. Testability which is the ability to check a system/component's airworthiness.
8. Designed failure-effect limits, including the capability to sustain damage that limits the safety effects of a failure.
9. Designed failure paths that control and direct the effects of a failure in a way that limits its safety impact.

10. Margins or factors of safety that allow for any undefined or unforeseeable adverse conditions.
11. Error tolerance that considers the adverse effects of foreseeable errors during the airplane's design, test, manufacture, operation, and maintenance.

5

Airplane and Appliance Certification

Introduction

United States law imposes upon the Administrator responsibility for safety that includes

> *...safe flight of civil aircraft in air commerce by prescribing...minimum standards required in the interest of safety for appliances and for the design, material, construction, quality of work, and performance of aircraft, aircraft engines, and propellers.*[1]

The certification of hardware[2] uses four processes: Type Certification, Production Certification, the Technical Standard Order system, and Parts Manufacturing Authorizations.

Certification is the responsibility of the Aircraft Certification Service through the Aircraft Certification Directorates (Figure 5-1). Each directorate is responsible for specific types of products. The Flight Standards Service through the Aircraft Evaluation Group participates in the process.[3]

New airplane projects

The development of a new airplane through certification takes 5 to 10 years. It begins with a product development stage. During this

[1]49 USC 44701.

[2]Software is included in this context.

[3]See Chapter 2 for an explanation of the responsibilities of these organizations.

Figure 5-1 *Certification Directorates.*

time (up to 5 years) the manufacturer commits business and technical resources to develop a configuration. This consists of numerous "paper" airplanes in response to a variety of inputs. The designs develop from market surveys, discussions with potential customers, regional and world economic demand forecasts, and evaluation of potential competitors. Available and near-term developing engineering and manufacturing technologies are studied for possible use in new airplanes. During this time such things as range and payload characteristics begin to evolve. The market is defined. Power plants are selected. Component, manufacturing vendors and risk-sharing partners are solicited and brought into the design. Eventually market demands and competitive pressures narrow the choices among these paper airplanes to a configuration, which it is believed will meet the future demands of that very limited airline customer base. Designing and building an airplane is an enormous financial gamble on the future of the economics of the airline industry and on the value of the design and technology for the potential 30-year life of the airplane.

As the preliminary design becomes firm, manufacturers begin discussions with the FAA to familiarize them with the impending project. This is advantageous to all parties. Lines of communication are opened. There is adequate time to define the project, brief the participants on the design, begin development of a certification plan, and identify issues. Both parties are better able to forecast budgets and structure resources. Issues will be resolved before they become problems.

Type Certification

Type Certification validates the conformity of the design to the basis of certification. It applies to airplanes, power plants, and propellers. The certification process is similar to Air Carrier Certification. It is divided into a series of steps. It involves a very large group of people. It requires up to 5 years for a new airplane design. The process is shown in Figure 5-2. FAA activity is focused through the Type Certification Board.

Type Certification Boards

Type Certification Boards (TCB) are established to acquaint the applicant and the FAA with the certification project, resolve significant problems, and establish a schedule for overall accomplishment of the type certification program. The Aircraft Certification Office (ACO) manager for a project serves as chair. Members include appropriate FAA management, engineering, flight test, and manufacturing people from within the ACO, National Resource Specialists, Aircraft Evaluation Group (AEG) representatives and of course representatives from the manufacturer. At least three board meeting are convened during the project: a preliminary, preflight, and final.

Type Certificate application

A preliminary Type Certification Board meeting is held when the manufacturer submits a Type Certificate application (Figure 5-3).

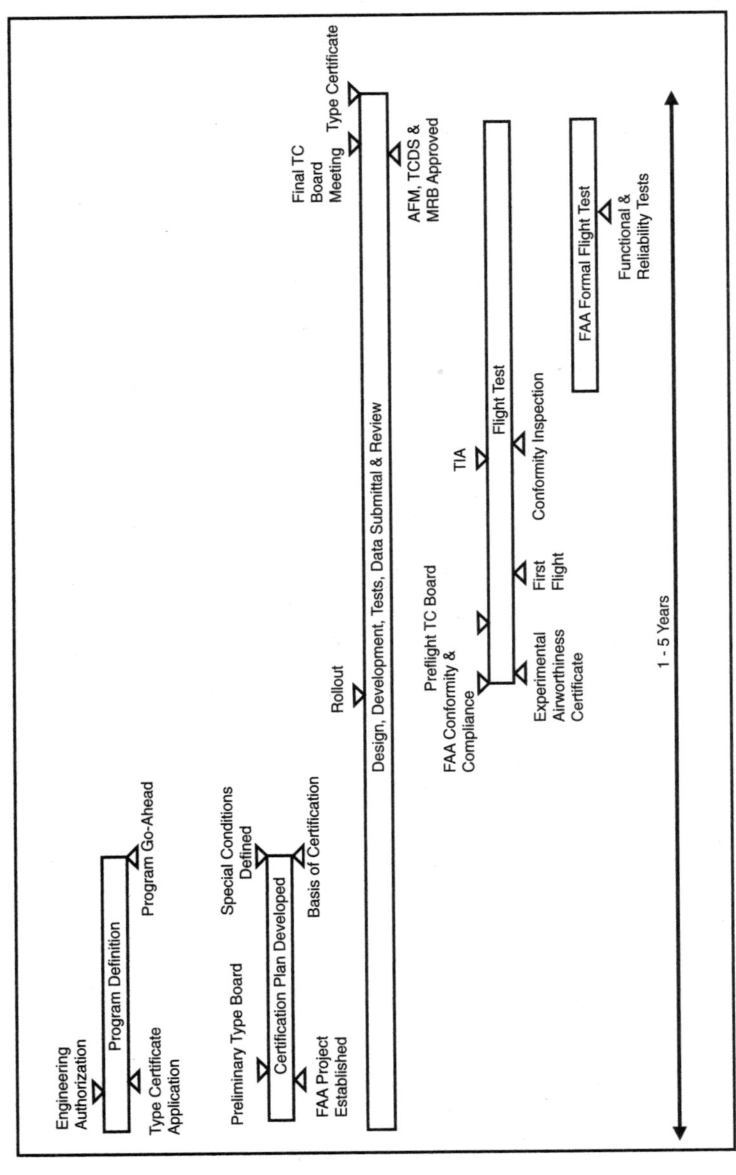

Figure 5-2 *Type Certification Process.*

BURNING STUMP INDUSTRIES, INC.

350 Waytofar North Blvd.
Iqaluit, Nunavut, XOA OHO
Canada

August 22, 2000

Department of Transportation
Federal Aviation Administration
FAA Building, King County International Airport
Seattle, WA 98108

Gentlemen:

Enclosed is a completed FAA Form 312, constituting an application for a Type Certificate for the 253 series aircraft.

Also enclosed is a copy of a three view drawing for the model.

FAR 21.17(b) notes that the application for Type Certification of a transport category aircraft is effective for five (5) years, unless an applicant shows at the time of application that his product requires a longer period of time for design, development and testing.

Due to the complexity of this program and because probably more than one version will be involved, it is anticipated that longer than five years will be required for the design, development and testing of this aircraft. Therefore, we request that the application for the 253 be extended to six (6) years.

A Preliminary Type Certification Board meeting should be held at an appropriate time in the development of the 253 aircraft. Our tentative schedule indicates this to be during the third quarter of 2000.

Sincerely,

/Original signed by/

Anton J. Gasnarfus
Airworthiness Compliance
Burning Stump Industries

Enclosures:

- **Involves the following steps**
  - Type Certificate application
  - Establish Basis of certification
  - Demonstrate conformity
    - Type design details
    - Testing
      - Structure
      - Systems
      - Flight test
    - Data analysis
    - Submission of Statement of Conformity
    - FAA review of substantiation/conformity
    - Award of Type Certificate
  - Details are addressed in FAA order 8110.4A, Type Certification Process

Figure 5-3 *Type Certification application.*

Informal discussions between the FAA and the manufacturer become formalized when the manufacturer submits this application.[4] Application is made as early as possible, long before a formal business commitment to the airplane. A clear understanding of the scope of the project and the ground rules and administrative procedures for the project must be established to the satisfaction of both parties to avoid later surprises or misunderstanding.

Significant issues are identified. Examples that must be substantially resolved before certification activity begins include:

- Definition of the Certification Basis.
- Determination of the acceptable means of compliance to be used by the manufacturer.
- Equivalent Safety Findings proposed or made.
- Identification of unsafe conditions, features, materials, etc., that could preclude certification.
- Identification of topics or items that are or could become controversial or may otherwise require resolution.
- Environmental considerations that must be satisfied prior to certification over and above Environmental Protection Agency (EPA) air pollution requirements and Noise Control Act environmental evaluations.
- Export requirements that include identifying the extent of FAA findings of compliance on behalf of a foreign civil airworthiness authority.
- Identification of rulemaking actions such as Special Conditions and exemption petitions.

During the meeting, the board is briefed on the project including details of the design. Lines of communication are established. The certification plan is discussed including schedules, milestones, and so on.

Certification plans

Certification plans are submitted early in the project. They contain:

- General information regarding the project, including such information as model designation, responsible organizations, and contacts.

[4] 14 CFR 21.15.

- A general description of the airplane, including sketches and schematics including the required three-view drawing.
- The proposed certification basis, including applicable Federal Aviation Regulation paragraphs and subparagraphs, exemptions, and special conditions.
- How compliance will be shown, such as ground test, flight test, analysis, similarity, and equivalent means of compliance, and what will be submitted to show compliance.
- A project schedule showing major milestones, preliminary hazard analysis submittal, detail submittals, when conformity inspection and testing are required, first flight, flight test, and when final certification is expected.
- Identification of all Designated Engineering Representatives (DERs), their specialties, and if the DER will be approving the data or recommending approval of the data.

Basis of Certification

During and immediately following the preliminary TCB meeting the first critical element is accomplished. It is the formalization of the Basis of Certification. The Basis of Certification bounds the project by defining the rules and conditions under which the design is to be certificated. It controls the activities of both parties. It defines what will determine airworthiness. Principal FARs[5] governing certification are identified. The Basis of Certification is published in the *Federal Register*. It is also included in the Type Certificate Data Sheet. The Basis of Certification (Figure 5-4) defines the airworthiness of the airplane. It is central to all maintenance activity.

Special conditions

Any unique features of the proposed design will be compared to existing regulations to identify those not covered in the FAR. The rules governing certification of these items are then defined by establishing

[5]Air carrier airplanes most commonly use 14 CFR 25 transport category airplanes. The governing FAR is identified through a given amendment number. They are usually the applicable regulations in effect on the date of the application.. This does not mean that the FAA may not subsequently impose a given amendment to the design. Rather, it means there is a consistent definition of the rules that does not burden the certification process by recurring routine FAR changes. It precludes continuing design changes during the life of the certification project.

[6]14 CFR 21.16.

Project:	The Boeing Company Model 767-200 Project No. CT1180NW-D	Item: Stage:	 3
Reg. Ref.:	FAR 21.17	Date:	June 2, 1962
National Policy Ref.	Federal Register Vol. 46 No. 5 Pages 2241-2243	Compliance Status: Issue Status:	 CLOSED
Subject:	Applicable Regulations	Branch Action	100/120/130 140/160/110 180
		Compliance Target:	Pre-TC

Statement of Issues: The proposed certification basis of the Boeing Model 767-200 is:

Federal Aviation Regulations (FAR) Part 25 Amendments 25-1 through 25-37.

Federal Aviation Regulations (FAR) Part 36 with Amendments 36-1 through 36-12.

Special Federal Aviation Regulation Part 27.

In addition to the above regulations, Boeing has elected to comply with the following later amendments to Part 25.

25.345 Amendment 46
25.351(a) Amendment 46
25.365(e)(1),(2) Amendment 54
25.629 Amendment 46
25.697 Amendment 46
27.733 Amendment 49
25.803 Amendment 46
25.901(d) Amendment 46
25.1103(a),(b)(2),(d),(e),(f) Amendment 46
25.1142 Amendment 46
25.1522 Amendment 46

Boeing has also elected to comply with Part 25, Amendments 25-38 through 25-45, Except portions of Amendment 25-38 (Sections 25.979(d) and (e) and Section 25.1143(e); Amendment 40 (Sections 25.901(b)(1)(i), 25.1091(e) and 25.1093(b)); Amendment 41 (Section 25.1438) and ; Amendment 42 (Section 25.109).

Special Conditions may be issued and amended, as necessary, as a part of the type certificate basis if the Administrator finds that the airworthiness standards do not contain adequate or appropriate safety standards because of novel or unusual design features of the Boeing Model 767-200 series airplane.

- **Preliminary type board meeting between applicant and the cognizant certification FAA region**
  - Design description
  - Certification standards for the design
  - Acceptable compliance methods
  - FAA interpretations
  - DER approval program
  - Maintenance considerations
  - Operations considerations
  - Determination of open issues
  - Special conditions
  - Exemptions
- Final version published in the *Federal Register*

Figure 5-4 *Basis of Certification.*

"special conditions."[6] A special condition (Figure 5-5) contains only such airworthiness standards as are necessary to establish a level of safety equivalent to that established by the applicable regulations. Special conditions are unique to the specific certification program in which they are issued. They are listed in the Basis of Certification.

Special conditions are not used to upgrade the airworthiness regulations when novel or unusual design features are not involved. Whenever the FAA determines that an upgrading of the airworthiness regulations is warranted, the upgrading must be accomplished through the rulemaking process. This prevents ad hoc regulation by using a special condition. Adherence to this principle is sometimes strained.

Issue papers

Issues focus on the regulatory compliance questions that arise during the project. The FAA uses a structured system for describing and tracking issues as they arise called *issue papers*. They provide visibility to all the participants and prevent substantive items from being lost or overlooked. They track the resolution of significant Type Certification questions and problems occurring throughout a program. Routine items relative to showing compliance are not normally raised as significant issues unless special problems are anticipated or develop during the course of the program.

An example of an issue is the Basis of Certification. New issue papers may be proposed to the TC Board at any time during the certification process. Issue papers passed through the preliminary TCB are assembled and published in an Issues Book. The book is used to track the issues until final resolution. All issues must be essentially closed before a Type Certificate is issued.

Equivalent level of safety finding

These findings are made when literal compliance with a certification regulation cannot be shown and compensating factors exist. Equivalent levels of safety findings are listed in the Basis of Certification.

Conformity

Once the Basis of Certification is established, the applicant for a Type Certificate must demonstrate the conformity of the design to the applicable regulations and special conditions imposed. Throughout this period as the design details are committed additional TCBs

[Federal Register: November 30, 1999 (Volume 64, Number 229)]
[Rules and Regulations]
[Page 66721-66723]

From the Federal Register Online via GPO Access [wais.access.gpo.gov]
[DOCID:fr30no99-5]

DEPARTMENT OF TRANSPORTATION Federal Aviation Administration

14 CFR Part 23

[Docket No. CE145; Special Conditions No. 23-096A-SC]

Special Conditions: Raytheon Model 390 Airplane

AGENCY: Federal Aviation Administration (FAA), DOT.

ACTION: Amended final special conditions; request for comments.
SUMMARY: This document amends special conditions issued to the Raytheon Aircraft Company for the Raytheon Model 390 airplane and requests comments on the revised portion of the amended special conditions. The Small Airplane Directorate issued final special conditions for this airplane on July 9, 1999, and published them on July 23, 1999 (64 FR 39899). The special conditions contained a requirement for operating limitations for weight and loading distribution already covered by an exemption issued to Raytheon Aircraft Company on December 12, 1996 (Exemption No. 6558, Docket No. 132CE). Accordingly, the portion of the special conditions that covers the operating limitations has been amended to remove the additional requirement. Only the revised sections are contained in this document. Additionally, the special condition for turning flight and accelerated turning stalls has been amended to include a power-at-idle condition. This condition is included to make these special conditions consistent with previously approved special conditions for a similar airplane.

DATES: The effective date of these special conditions is November 15, 1999. Comments must be received on or before December 30, 1999. ADDRESSES: Comments on these special conditions may be mailed in duplicate to: Federal Aviation Administration, Regional Counsel, ACE-7, Attention: Rules Docket CE145, 901 Locust, Room 506, Kansas City, Missouri 64106; or delivered in duplicate to the Regional Counsel at the above address.

Figure 5-5 *Typical special condition.*

are held along with design reviews. Detailed engineering analyses, laboratory test data, and demonstrations of structure, systems, and components are submitted to the FAA for review and approval.

Flight test

As the design progresses and becomes firm the first airplane is manufactured. For a new design, this can be up to 3 to 4 years after commitment of the project. A significant part of demonstrating conformity is the conduct of various flight tests. Before the initiation of a flight test, the airplane is subjected first to a conformity check. When the results of this test are accepted the FAA issues a Type Inspection Authorization authorizing the first flight. This is a big event. A formal TCB is convened to review the data. The manufacturer, to verify flying qualities, determine drag levels and crosswind capability, and so forth, does the initial test flying. This initial flight test data are reviewed and a conformity inspection is conducted to verify the airplane configuration, a TCB is held, and the formal FAA flight test program begins. Flight test and other certification work continues right up to the day of Type Certification.

Type Certificates

When full conformity to the Basis of Certification has been demonstrated the design is awarded a Type Certificate. A Type Certificate (Figure 5-6) includes the type design, the operating limitations, the Type Certificate Data Sheet (TCDS), the applicable regulations, and any other conditions or limitations prescribed by the Administrator.[7] The certificate consists of:

- The Type Design which is the entire body of data including the drawing system used to define the airplane
- The Basis of Certification which defines the rules governing the certification that define the airworthiness of the design
- Substantiation, which is all of the conformity, flight test data, inspection results and other documentation
- The approved Airplane Flight Manual (AFM) which contains the operating limitations imposed upon the type design
- The Type Certification Data Sheet

Type Certificates contain proprietary information. Consequently, all of the data are not made available to the public domain. Type Certificates are transferable. The holder of the Type Certificate may obtain Airworthiness Certificates, a Production Certificate, and approval of replacement parts. Type Certificates are transferable.

Provisional Type Certificate

Provisional Type Certificates are issued when all requirements for a Type Certificate are not complete but the applicant can show that

[7] 14 CFR 21.41.

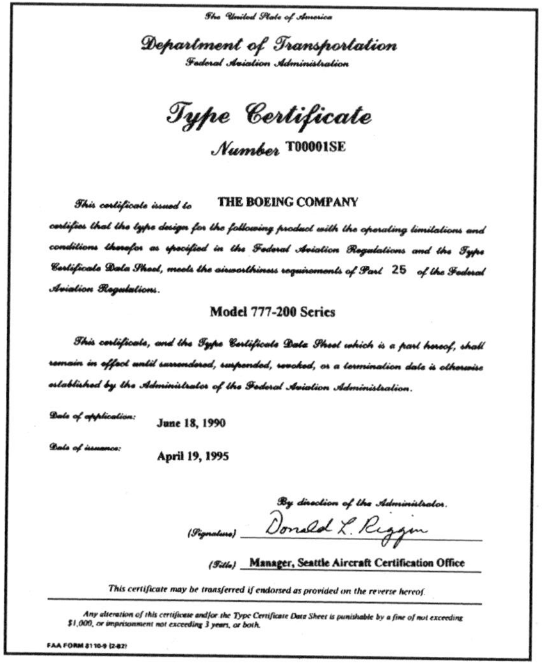

The United State of America

Department of Transportation

Federal Aviation Administration

Type Certificate

Number T00001SE

This certificate issued to THE BOEING COMPANY certifies that the type design for the following product with the operating limitations and conditions therefor as specified in the Federal Aviation Regulations and the Type Certificate Data Sheet, meets the airworthiness requirements of Part 25 of the Federal Aviation Regulations.

Model 777-200 Series

This certificate, and the Type Certificate Data Sheet which is a part hereof, shall remain in effect until surrendered, suspended, revoked, or a termination date is otherwise established by the Administrator of the Federal Aviation Administration.

Date of application: June 18, 1990

Date of issuance: April 19, 1995

By direction of the Administrator.

(Signature) Donald L. Riggin

(Title) Manager, Seattle Aircraft Certification Office

This certificate may be transferred if endorsed as provided on the reverse hereof.

Any alteration of this certificate and/or the Type Certificate Data Sheet is punishable by a fine of not exceeding $1,000, or imprisonment not exceeding 3 years, or both.

FAA FORM 8110-9 (2-82)

- **Is transferable and considered proprietary to the applicant**
- **Consists of**
  - The type design
  - The Basis of Certification
  - Substantiation
  - The operating limitations (AFM)
  - The Type Certification Data Sheet
- **The holder of the Type Certificate may**
  - Obtain Airworthiness Certificates
  - Obtain a Production Certificate
  - Obtain approval of replacement parts
- **Variations**
  - Provisional Type Certificates
  - Surplus military aircraft
  - Restricted category aircraft
  - Import products

Figure 5-6 *Type Certificate.*

there is no feature, characteristic, or condition that would make the aircraft unsafe when operated in accordance with appropriate limitations. Provisional Type Certificates are not transferable.

A Provisional Type Certificate is appropriate and necessary for the following aircraft uses:

- Flight crew training
- Demonstration flights by the manufacturer for prospective purchasers
- Market surveys by the manufacturer
- Flight checking of instruments, accessories, and equipment
- Service testing of the aircraft[8]

Changes to the Type Certificate

Throughout the life of the airplane new design features are incorporated into the model. All require certification. The majority of the changes that occur are simple straightforward inclusion of additional or customized features. Type Certificates are amended when the change to the type design is not so extensive as to require a new certificate. This is done in the manner of original certification. Existing certification data are amended to incorporate these changes. When changing the Type Certificate only those items or features in the design that have not previously been certificated must go through the process.

As the number of model variants increases the latest version frequently bears little resemblance to the original design except for its cosmetic appearance and basic structure. Systems and power plants amass numerous changes with the passage of time. How extensive must these changes become before a new Type Certificate is required? The regulations[9] address this but are ambiguous. This is advantageous to all parties of a certification. Full Type Certification projects are inordinately costly undertakings. Unless there is true demonstrated need, a new Type Certificate project is avoided. However, determining the point at which a new Type Certificate project must be initiated is an interesting debate that is sometimes contentious but not rancorous. It is a debate with merit.

[8]See 14 CFR 21, Subpart C; 14 CFR 91.317; and 14 CFR 121.207 for regulations of issuance and operation limitations.

[9]14 CFR 21.19 and 21.93.

Avoiding a new Type Certificate has another advantage. I call it the avoidance of the "I'll never agree to that again!" syndrome of Type Certification. Any large project is subject to hindsight. New airplanes are no different. After the project is complete, all parties have victories, losses, and regrets with various elements of the undertaking. This leads to individual and collective evaluations of the actions and strategies followed. In the losses and regrets column each party at some point will say, "I'll never do that again!" Opening a new Type Certification project against an extensively changed type design "opens old wounds" and is counterproductive to both sides.

Don't misinterpret my words. I am not implying that avoiding a new Type Certificate for a model variant compromises safety. None of the parties to the process will allow that to happen.

Type Certificate Data Sheet

The Type Certificate Data Sheet (TCDS) (Figure 5-7) documents the conditions and limitations necessary to meet the airworthiness requirements. It is a highly structured reference document maintained by the Type Certificate holder and published by the FAA.

Among the items contained in the list are the Basis of Certification, airplane serial numbers eligible, allowable powerplants, limiting gross weights, approved installed equipment, and weight and balance information.

Type Certification Data Sheets are available by subscription from the FAA. They are also accessible on the Web.

Supplemental Type Certificate

Any individual may change the type design. A Supplemental Type Certificate (STC) (Figure 5-8) is issued for major design changes to a Type Certificate when the change is not so extensive as to require a new Type Certificate. Minor changes do not require an STC. Type Certificate certification projects may be very small or very involved. Typical examples of STCs are the installation of a new passenger interior, a new powerplant type, or conversion of the airplane from a passenger to a freight airplane by installing a main deck cargo door and cargo handling system.

Supplemental Type Certificates are frequently very restrictive, limiting the design change to specific airplane serial numbers rather than a complete model series. An STC like a TC is transferable.

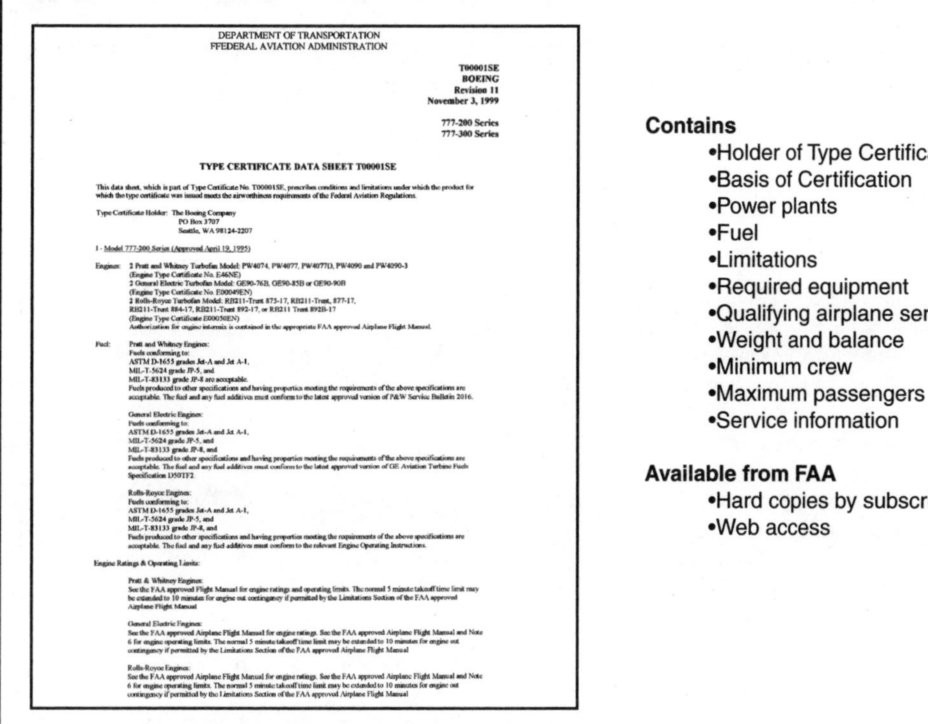

DEPARTMENT OF TRANSPORTATION
FFEDERAL AVIATION ADMINISTRATION

T00001SE
BOEING
Revision 11
November 3, 1999

777-200 Series
777-300 Series

TYPE CERTIFICATE DATA SHEET T00001SE

This data sheet, which is part of Type Certificate No. T00001SE, prescribes conditions and limitations under which the product for which the type certificate was issued meets the airworthiness requirements of the Federal Aviation Regulations.

Type Certificate Holder: The Boeing Company
PO Box 3707
Seattle, WA 98124-2207

I - Model 777-200 Series (Approved April 19, 1995)

Engines: 2 Pratt and Whitney Turbofan Model: PW4074, PW4077, PW4077D, PW4090 and PW4090-3
(Engine Type Certificate No. E46NE)
2 General Electric Turbofan Model: GE90-76B, GE90-85B or GE90-90B
(Engine Type Certificate No. E00049EN)
2 Rolls-Royce Turbofan Model: RB211-Trent 875-17, RB211-Trent, 877-17, RB211-Trent 884-17, RB211-Trent 892-17, or RB211 Trent 892B-17
(Engine Type Certificate E00050EN)
Authorization for engine intermix is contained in the appropriate FAA approved Airplane Flight Manual.

Fuel: Pratt and Whitney Engines:
Fuels conforming to:
ASTM D-1655 grades Jet-A and Jet A-1,
MIL-T-5624 grade JP-5, and
MIL-T-83133 grade JP-8 are acceptable.
Fuels produced to other specifications and having properties meeting the requirements of the above specifications are acceptable. The fuel and any fuel additives must conform to the latest approved version of P&W Service Bulletin 2016.

General Electric Engines:
Fuels conforming to:
ASTM D-1655 grades Jet-A and Jet A-1,
MIL-T-5624 grade JP-5, and
MIL-T-83133 grade JP-8, and
Fuels produced to other specifications and having properties meeting the requirements of the above specifications are acceptable. The fuel and any fuel additives must conform to the latest approved version of GE Aviation Turbine Fuels Specification D50TF2

Rolls-Royce Engines:
Fuels conforming to:
ASTM D-1655 grades Jet-A and Jet A-1,
MIL-T-5624 grade JP-5, and
MIL-T-83133 grade JP-8, and
Fuels produced to other specifications and having properties meeting the requirements of the above specifications are acceptable. The fuel and any fuel additives must conform to the relevant Engine Operating Instructions.

Engine Ratings & Operating Limits:

Pratt & Whitney Engines:
See the FAA approved Flight Manual for engine ratings and operating limits. The normal 5 minute takeoff time limit may be extended to 10 minutes for engine out contingency if permitted by the Limitations Section of the FAA approved Airplane Flight Manual

General Electric Engines:
See the FAA approved Airplane Flight Manual for engine ratings. See the FAA approved Airplane Flight Manual and Note 6 for engine operating limits. The normal 5 minute takeoff time limit may be extended to 10 minutes for engine out contingency if permitted by the Limitations Section of the FAA approved Airplane Flight Manual

Rolls-Royce Engines:
See the FAA approved Airplane Flight Manual for engine ratings. See the FAA approved Airplane Flight Manual and Note 6 for engine operating limits. The normal 5 minute takeoff time limit may be extended to 10 minutes for engine out contingency if permitted by the Limitations Section of the FAA approved Airplane Flight Manual

Figure 5-7 *Type Certification Data Sheet.*

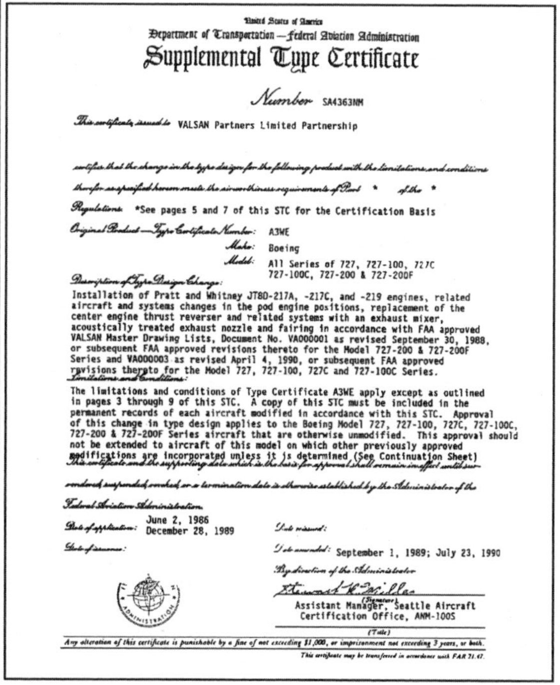

United States of America
Department of Transportation—Federal Aviation Administration

Supplemental Type Certificate

Number SA4363NM

This certificate, issued to VALSAN Partners Limited Partnership

certifies that the change in the type design for the following product with the limitations and conditions therefor as specified hereon meets the airworthiness requirements of Part * of the * Regulations. *See pages 5 and 7 of this STC for the Certification Basis

Original Product—Type Certificate Number: A3WE
Make: Boeing
Model: All Series of 727, 727-100, 727C 727-100C, 727-200 & 727-200F

Description of Type Design Change:
Installation of Pratt and Whitney JT8D-217A, -217C, and -219 engines, related aircraft and systems changes in the pod engine positions, replacement of the center engine thrust reverser and related systems with an exhaust mixer, acoustically treated exhaust nozzle and fairing in accordance with FAA approved VALSAN Master Drawing Lists, Document No. VA000001 as revised September 30, 1988, or subsequent FAA approved revisions thereto for the Model 727-200 & 727-200F Series and VA000003 as revised April 4, 1990, or subsequent FAA approved revisions thereto for the Model 727, 727-100, 727C and 727-100C Series.

Limitations and Conditions:
The limitations and conditions of Type Certificate A3WE apply except as outlined in pages 3 through 9 of this STC. A copy of this STC must be included in the permanent records of each aircraft modified in accordance with this STC. Approval of this change in type design applies to the Boeing Model 727, 727-100, 727C, 727-100C, 727-200 & 727-200F Series aircraft that are otherwise unmodified. This approval should not be extended to aircraft of this model on which other previously approved modifications are incorporated unless it is determined (See Continuation Sheet)

This certificate and the supporting data which is the basis for approval shall remain in effect until surrendered, suspended, revoked, or a termination date is otherwise established by the Administrator of the Federal Aviation Administration.

Date of application: June 2, 1986 December 28, 1989
Date reissued:
Date of issuance:
Date amended: September 1, 1989; July 23, 1990

By direction of the Administrator

(Signature)
Assistant Manager, Seattle Aircraft Certification Office, ANM-100S
(Title)

Any alteration of this certificate is punishable by a fine of not exceeding $1,000, or imprisonment not exceeding 3 years, or both.

This certificate may be transferred in accordance with FAR 21.47.

- Are transferable
- Are considered proprietary
- Supplement an existing Type Certificate
- Address a specific design feature
  - Examples: new cargo loading system or a new engine type
- Frequently are limited to specific airplane serial numbers
- An Advisory Circular provides an index
- FAA Notice 8110.69, Supplemental Type Certificate Requirements, contains additional information

Figure 5-8 *Supplemental Type Certificate.*

Designated Alteration Station

Some domestic repair stations, air carriers, and manufacturers may hold a Designated Alteration Station (DAS) authorization. A DAS authorization holder has, within certain limits, the authority to issue Supplemental Type Certificates. The designation includes authority to issue experimental certificates of airworthiness for aircraft undergoing supplemental type certification and the issuance of amended standard airworthiness certificates for aircraft which incorporate an STC issued by the DAS.

Production Certification

Production Certification is similar to Type Certification. An FAA Production Certification Board oversees the process. It usually is done concurrently with the Type Certification. It consists of a demonstration of production conformity showing that the production facility is capable of manufacturing the product to the approved type design. Certification includes an evaluation of the manufacturing facilities, administrative practices, production organization, personnel qualifications, and quality system. Production Certificates are issued to either the holder of a Type Certificate, an STC, or a licensee. Production Certificates are not transferable.

At the heart of production is a quality control system that controls all elements of manufacture, including control over all suppliers.

Production Certificates

Once the Production Certification requirements have been met the manufacturer is awarded a Production Certificate (Figure 5-9). The certificate attests that the manufacturer has established and maintains a quality control system to assure that each airplane meets the design provisions of the Type Certificate. It authorizes the production of duplicate airplanes, engines, propellers, or appliances. Production Certificates are dependent upon a physical facility and administrative system. Consequently, they are not transferable.

Production Limitation Record

The Production Certificate may not always authorize production of every model listed on the Type Certificate. The Production Limitation Record (PLR) lists each specific product. The Production

The United States of America
Department of Transportation
Federal Aviation Administration

Production Certificate

Number 700

This certificate issued to
THE BOEING COMPANY
whose business address is
7755 East Marginal Way South
Seattle, Washington 98108

and whose manufacturing facilities are located at
Please see Supplemental Sheet for addresses

authorizes the production, at the facilities listed above, of reasonable duplicates of AIRPLANES, AND TYPE DESIGN CHANGES APPROVED BY SUPPLEMENTAL TYPE CERTIFICATES *which are manufactured in conformity with authenticated data, including drawings, for which Type Certificates specified in the pertinent and currently effective Production Limitation Record were issued. The facilities, methods, and procedures of this manufacturer were demonstrated as being adequate for the production of such duplicates on date of* June 13, 1949, and as amended.

Duration: *This certificate shall continue in effect indefinitely, provided, the manufacturer continuously complies with the requirements for original issuance of certificate, or until the certificate is canceled, suspended, or revoked.*

Date issued:
June 13, 1949
This amendment - ~~October 23, 1985~~

By direction of the Administrator
H.E. Waterman
Chief Engineering and Manufacturing Division
Manager, Seattle Aircraft Certification Office

This Certificate is not Transferable, AND ANY MAJOR CHANGE IN THE BASIC FACILITIES, OR IN THE LOCATION THEREOF, SHALL BE IMMEDIATELY REPORTED TO THE APPROPRIATE REGIONAL OFFICE OF THE FEDERAL AVIATION ADMINISTRATION.

Any alteration of this certificate is punishable by a fine of not exceeding $1,000, or imprisonment not exceeding 3 years or both

FAA FORM 8120-4 (12-69) SUPERSEDES FAA FORM 333

- Issued to the holder of a Type Certificate, Supplemental Type Certificate, or licensee
- Requires
  - An administrative system
  - Drawing system
  - Quality control system
    - Materials Review Board (MRB)
    - Inspection program
    - Vendor control
  - Physical facility
- Production Certificates are not transferable
- Authorizes the manufacture of duplicates

Figure 5-9 *Production Certificate.*

- Organization and responsibilities including management and inspectors
- Design data control
- Software control
- Manufacturing Processes including special processes
- Statistical quality control
- Tool and gage control
- Testing including NDT
- Supplier Control
- Nonconforming material
- Material handling and storage
- Airworthiness determination
- FAA reporting requirements
- Internal audit
- Global manufacturing
- Manufacturer's maintenance facility

Figure 5-9 *(Continued) Production Certification Topics.*

Limitation Record lists each model authorized for manufacture under the Production Certificate, and the date of such authorization.

Amendment of the Production Certificate

Amendments to a Production Certificate cover only changes to the existing quality control data when production of a new product involves changes in the quality control system. This includes, for example, installing new tooling or manufacturing processes.

Airworthiness Certification

Each completed production airplane is subjected to a final inspection for conformity to the defined configuration including freedom from damage and/or contamination, and safe operating condition.

This is normally accomplished by Designated Manufacturing Inspection Representatives (DMIR). The work includes determining that:

- The empty weight and center of gravity are accurate, including applicable loading charts and instructions.
- The aircraft equipment list is accurate.
- Required functional tests meet the approved design provisions.
- Each aircraft is subjected to a flight test.

The Technical Standard Order system

Many devices or parts are capable of being used on more than one airplane. It is impractical and costly to require that a brake, for instance, be certified every time it is installed on a new airplane. The Technical Standard Order (TSO) system[10] provides the means for certifying a common device once. It treats appliances as unique entities isolating them from the type design. Examples of items so certified are shown in Figure 5-10.

The advantages are evident. You certify devices only once. When a TSO-qualified device is included in the type design only substantiation of the installation is required during Type Certification. The appliance itself is already certified.

Technical Standard Orders (TSOs) are developed, controlled, and published by the Aircraft Certification Service. Draft TSOs and revisions are made available for comment before their adoption. The standards[11] used for TSO certification of devices are normally defined by accepted industry standards established by such technical organizations as the Radio Technical Commission for Aeronautics (RTCA), Society of Automotive Engineers (SAE).

Qualification of a device under a TSO is similar to Type and Production Certification but on a much smaller scale. Substantiation data that demonstrate conformity and production capability are submitted to the FAA for review and approval. Once approved the FAA issues a TSO Authorization. This authorization is the design

[10] 14 CFR 21 Subpart O.

[11] Current Technical Standard Orders are indexed in AC 20-110J, Index of Aviation Technical Standard Orders. This Advisory Circular also provides advice for obtaining copies of a given TSO.

Department of Transportation
Federal Aviation Administration
Aircraft Certification Service
Washington, DC

TSO-C149

Date: 4/24/98

Technical Standard Order

Subject: TSO-C149, AIRCRAFT BEARINGS

1. **PURPOSE.** This technical standard order (TSO) prescribes property test requirements to obtain the minimum performance of aircraft bearings to be identified with the applicable TSO marking.

2. **APPLICABILITY.** The standards of this TSO apply to the types of bearings described in appendix 1, Aircraft Bearing Property Test Requirements, intended for rotation and/or oscillatory applications in the manufacture and maintenance of aircraft products. The standards of this TSO are also adaptable to manufacturer's catalog bearings and bearings of proprietary designs. This TSO shall not be used for standard parts or parts known to be used in critical applications.

3. **REQUIREMENTS.** Aircraft bearings that are to be identified with this TSO and that are manufactured on or after the date of this TSO must meet the minimum performance standards specified in the manufacturer's part drawing(s) and applicable part specification(s) submitted with the bearing manufacturer's application for TSO authorization.

a. Test Requirements. The required performance shall be demonstrated by accomplishing the tests specified for each property in the part drawing(s) and applicable part specification(s), in accordance with the test procedures specified in appendix 1.

b. Deviations. Alternative test procedures or analytical data that produce an equivalent level of safety may be used if specified at the time of TSO application and approved in accordance with 14 CFR §21.609.

4. **MARKING.**

a. In addition to the marking specified in 14 CFR §21.607(d), the bearing type, the lubrication date (if applicable), and the manufacturer's inspection lot number shall be permanently and legibly marked on each package or container.

b. Each individual bearing that is manufactured under this TSO must be permanently and legibly marked with at least the name or symbol of the manufacturer, the manufacturer's part number, and TSO number. When this is not practical, marking may be accomplished in a manner acceptable by the Administrator.

5. **DATA REQUIREMENTS.**

a. In accordance with 14 CFR §21.605(a) the following data must be furnished to the Aircraft Certification Office (ACO) manager having purview of the manufacturer's facility with each TSO application:

DISTRIBUTION: ZVS-326;A-W(IR)-3;A-X(FS)-3;A-X(CD)-4;
A-FFS-1,2,7,8(LTD);A-FAC-0(MAX);AVN-1 (2 cys)

IR Form 8150-1

- Individual components may be certificated
  - As a part of the type design
    - Valid only when installed on the type design
  - Under the Technical Standard Order system defined in FAR 21
- Procedures for establishing airworthiness
- Establishes minimum performance standards for
  - Materials
  - Parts
  - Appliances
  - Processes
- Examples
  - Instruments
  - Wheels/tires/brakes
  - seats/seat belts
  - Communication/navigation equipment

Figure 5-10 *Technical Standard Order.*

and production approval issued to the manufacturer. It is not transferable. Changes to the device or the manufacturing operations must be documented and the device requalified by amendment to the TSO Authorization. TSO Authorizations are not transferable inasmuch as they contain production authorization.

Parts Manufacturing Authorizations (PMA)

Parts Manufacturer Approvals (Figure 5-11) are issued for the production of modification parts or replacement parts that may be installed on certain type certificated products. PMAs are not required for:

- Parts produced under a Type or Production Certificate.
- Parts produced by owners or operators for maintaining or altering their own products. This includes air carriers.
- Parts produced under a Technical Standard Order.
- Standard parts (such as bolts and nuts) conforming to established industry or U.S. specifications.

Repair stations or air carriers may manufacture replacement parts as part of their maintenance program without a PMA. These parts are acceptable, provided they are manufactured according to acceptable FAA-approved data.

Bilateral Aviation Safety Agreements

Bilateral Aviation Safety Agreements (BASA) are intended to facilitate reciprocal airworthiness certification of civil aeronautical products imported or exported between two signatory countries. They consist of an Executive Agreement, signed by the Department of State and its foreign counterpart, and the FAA and its foreign civil aviation counterpart.

They address implementation procedures for recognizing each other's airworthiness, maintenance, flight simulators, and environmental approvals.

Reportable items

Manufacturers must report[12] any failure, malfunction, or defect in their products that resulted or could result in any of the occurrences

[12] 14 CFR 21.3.

EXAMPLE - FAA-PMA LETTER
DEPARTMENT OF TRANSPORTATION
FEDERAL AVIATION ADMINISTRATION
Regional Office
601 East 12th Street
Kansas City, Missouri 64106

December 1, 1978

Aero-Parts, Incorporated
3212 Newton Street
St. Louis, Missouri

FEDERAL AVIATION ADMINISTRATION - PARTS MANUFACTURER APPROVAL

In accordance with the provisions of FAR 21, Subpart K, the FAA has found that the design data, as submitted by Aero-Parts, Inc., on September 16, 1977, meets the airworthiness requirements of the FARs applicable to the product(s) on which the part(s) is to be installed. Additionally, the FAA has determined that Aero-Parts, Inc., has established the fabrication inspection system required by FAR 21.303(h) at 3212 Newton Street, St. Louis, Missouri. Accordingly, Parts Manufacturer Approval (PMA) is hereby granted to Aero-Parts, Inc., to produce the replacement parts (or modification parts, as applicable) listed in the enclosed supplement(s) in conformity with the FAA-approved design data. Any subsequent changes to these design data must be approved in a manner acceptable to the FAA.

The following terms and conditions are applicable to this approval:

1. Aero-Parts, Inc., fabrication inspection system, methods, procedures and manufacturing facilities, including your suppliers, are subject to FAA surveillance or investigations. Accordingly, Aero-Parts, Inc., must advise their suppliers that their facilities are also subject to FAA surveillance and investigation.

2. Aero-Parts, Inc. must notify our district office (address) in writing within 10 days from the date the manufacturing facilities at which parts are manufactured are relocated or expanded to include additional facilities at other locations. This requirement also applies to Aero-Parts' suppliers, but only those who have been delegated major inspection authorization and those who furnish parts or related services where a determination as to safety and conformance to the approved design cannot or will not be made upon receipt at the approved receiving facility.

3. Aero-Parts, Inc., must make available to FAA, upon request, any pertinent information concerning their suppliers who furnish parts/services, including:

a. A description of the part or service;
b. Where and by whom the part or service will undergo inspection;
c. Any delegation of inspection duties;

Page 1

- Similar to a Production Certificate they are defined in FAR 21.303:
  - Limited to a specific location
  - Like a Production Certificate, the holder may use subcontractors subject to the limitations of the PMA
- Authorizes the manufacturer of a specific part which may be used on another certificated product:
  - Example: Replacement parts for type certificated products
- In recent years has been subject of considerable interest regarding:
  - Proper documentation
  - Definition of "standard parts"
  - "OEM" parts
  - "Bogus" parts
- FAA Order 8110.42, Parts Manufacturer Approval Procedures, addresses details.

Figure 5-11 *Parts Manufacturing Authorization.*

listed in Figure 5-12. This can include certain foreign manufacturers. Reports are made to the local Aircraft Certification Office within 24 hours[13] after it has been determined that a failure, malfunction, or defect has occurred.

Reported items are grouped with similar reports submitted by air carriers and published as Operational Difficulty Reports (ODR) (Figure 5-13). These reports are available on the Web as well as through a hard-copy subscription service.

Reporting is not required for instances of improper maintenance or usage that were reported to the FAA by another person under the Federal Aviation Regulations, or that have been reported to the National Transportation Safety Board.[14]

Certification documents

The manufacturer must at aircraft delivery provide:

- A valid Airworthiness Certificate
- A current approved Airplane Flight Manual
- A current weight and balance statement
- Instructions for continuing airworthiness
- Compliance status of ADs[15]

Airworthiness Certificates

Civil aircraft must have an Airworthiness Certificate (Figure 5-14) before they may be operated. They exist in several variants to meet numerous airplane types, configurations, and uses.

Standard Airworthiness Certificates

The standard airworthiness certificate (Figure 5-15) is by far the most common. They are issued for aircraft type certificated in the normal, utility, acrobatic, commuter, or transport category, and for manned free balloons. Also included are aircraft designated by the Administrator as special classes.

An Airworthiness Certificate is issued when the FAA finds that the aircraft fully conforms to the airworthiness requirements under

[13] Except weekends and holidays.

[14] 49 CFR 830.

[15] 14 CFR 91.417(a)(2)(v).

Fires caused by a system or equipment failure, malfunction, or defect
An engine exhaust system failure, malfunction, or defect that causes damage to the engine, adjacent aircraft structure, equipment, or components
The accumulation or circulation of toxic or noxious gases in the crew compartment or passenger cabin
A malfunction, failure, or defect of a propeller control system
A propeller or rotorcraft hub or blade structural failure
Flammable fluid leakage in areas where an ignition source normally exists
A brake system failure caused by structural or material failure during operation
A significant aircraft primary structural defect or failure caused by any autogenous condition (fatigue, under strength, corrosion, etc.)
Any abnormal vibration or buffeting caused by a structural or system malfunction, defect, or failure
An engine failure
Any structural or flight control system malfunction, defect, or failure which causes an interference with normal control of the aircraft for which derogates the flying qualities
A complete loss of more than one electrical power–generating system or hydraulic power system during a given operation of the aircraft
A failure or malfunction of more than one attitude, air speed, or altitude instrument during a given operation of the aircraft

Figure 5-12 *Reportable items.*

which it is certificated.[16] The presence of a Standard Airworthiness Certificate aboard the airplane is evidence that it is part of the airplane Type Certificate and was produced under a Production Certificate for that type design. It is not de facto evidence that the airplane is currently airworthy. The certificate is only valid as long as the airplane conforms to its Basis of Certification. There must be evidence that the airplane has had the required maintenance performed including applicable Airworthiness Directive accomplishment in order to keep it in conformity.

Special Airworthiness Certificates

Special Airworthiness Certificates (Figure 5-16) apply to all nonstandard aircraft.[17] Certificates in the restricted, special flight permit and experimental classifications are issued only for specific purposes. Operating limitations are generally prescribed with these types of certificates.[18]

[16] 14 CFR 23 through 31, as applicable, for airplanes, rotorcraft, and balloons, and in 14 CFR 21.23 for gliders.

[17] As classified under 14 CFR 21.175(b). They are issued when an aircraft meets the requirements of 14 CFR 21.185 through 21.199, as applicable for the certificate requested.

[18] 14 CFR 91.39 through 91.42, as applicable, specify limitations which must be complied with in all cases; however, the FAA inspector issuing the special Airworthiness Certificate may prescribe additional limitations if considered necessary for safety.

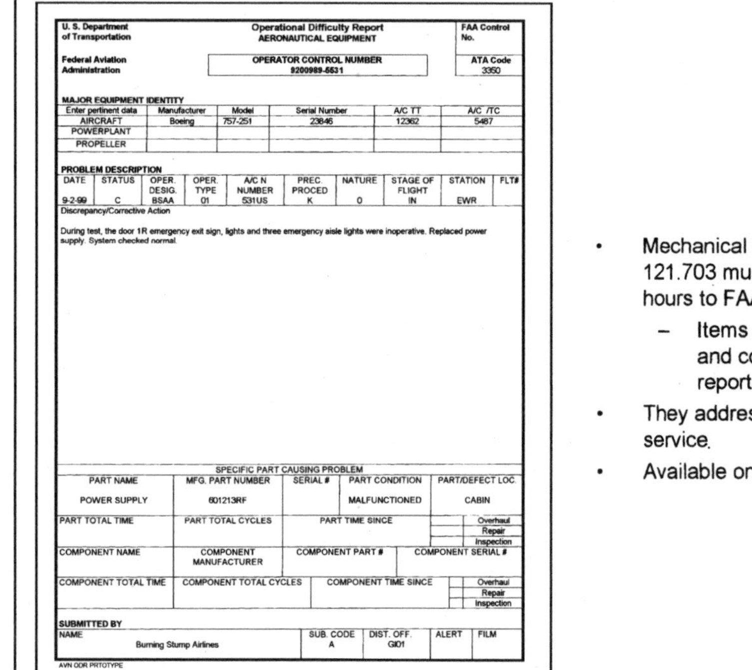

Figure 5-13 *Operational Difficulty Report.*

Special flight permits

Special flight permits[19] are issued for aircraft that may not currently meet applicable airworthiness requirements, but are capable of safe flight. They are normally only valid for one flight. They always contain specific limitations on the configuration of the airplane, the manner in which it may be operated, and meteorological conditions, and may include limitations on the route to be flown.

Although special flight permits may be issued only by the FAA, air carriers may be granted blanket authorization for ferry flight with one engine inoperative.[20] The authorization with appropriate limitations will be listed in the Operations Specifications.

Boeing and Airbus offer ferry flight opinions to their customers that may be used when applying for a ferry permit. However, manufacturer's opinions are not coordinated with the FAA nor do they constitute approval for the issuance of a special flight permit.

The approved Airplane Flight Manual (AFM)

An Airplane Flight Manual is furnished with each airplane. It may be published in paper or computerized versions. Each manual Log of Pages lists the specific airplane serial numbers applicable. It must be aboard the airplane.[21] However, air carriers may carry either the AFM or appropriate material from it in the Flight Operations Manual,[22] provided AFM data are clearly identified.

Furthermore, the operator may revise the AFM operating procedures sections and the presentation of performance data from them if the revised operating procedures and modified performance data presentation are (1) approved by the Administrator, and (2) clearly identified as Airplane Flight Manual requirements. This is the most common method of disseminating AFM data.

[19]Commonly called ferry permits.

[20]14 FR 91.611.

[21]14 CFR 121.141.

[22]One of several manuals used by an operator to meet the requirements of 14 CFR 121.133.

Standard Airworthiness Certificate (14CFR 21.183)	
Issued to normal, utility, acrobatic, commuter, and transport category aircraft; manned free balloons; and special classes of aircraft. It certifies that as of the date of issue the aircraft has been inspected and conforms to the Type Certificate, is in a condition for safe operation and conforms as applicable to the appropriate requirements of 14 CFR 34 and 36. Issued to one aircraft only. Must be prominently displayed. Valid only as long as the airplane is maintained to keep it in conformity. Import aircraft that meet appropriate criteria are also eligible.	
Export Airworthiness Certificate	
Substantiates a Class I aeronautical product as conforming to the Basis of Certification and thus capable of being awarded a Standard Airworthiness Certificate. Class I products are type certificated items.	
Special Airworthiness Certificate	
Experimental	Research and development; demonstrate compliance; crew training; exhibition; air racing; market survey; amateur built. 14 CFR 21.191, 21.193, and 21.195
Restricted	For airplanes certificated in the restricted category. Examples include military aircraft modified for special flight applications such as agricultural (spraying, dusting, and seeding, and livestock and predatory animal control); forest and wildlife conservation; aerial surveying (photography, mapping, and oil and mineral exploration); patrolling (pipelines, power lines, and canals); weather control (cloud seeding); aerial advertising (skywriting, banner towing, airborne signs, and public address systems). 14 CFR 21.185

Special Flight Permit	Movement for repairs, maintenance, or alterations; delivery or export of an airplane; production flight testing; evacuating from danger; customer demonstration flight; overweight operations. 14 CFR 21.197 and 21.199.
Limited	Applies to airplanes awarded a Limited Category Type Certificate. 14 CFR 21.189
Provisional	Applies to airplanes awarded a Provisional Type Certificate 14 CFR 21, Subpart C and Subpart I
Primary	Applies to airplanes awarded a Primary Category Type Certificate. These are type designs sold as kits rather than assembled airplanes. 14CFR 21.84
Airworthiness Approval Tags	
Export airworthiness approvals are issued as Form 8130-3 for Class II and III products. They are also used to substantiate conformity of domestically produced parts. They are also used to return to service parts having undergone overhaul or repair. Class II products are major components of Class I products. Class III products are any aeronautical products not Class I or Class II products.	

Figure 5-14 *Airworthiness Certificate summary.*

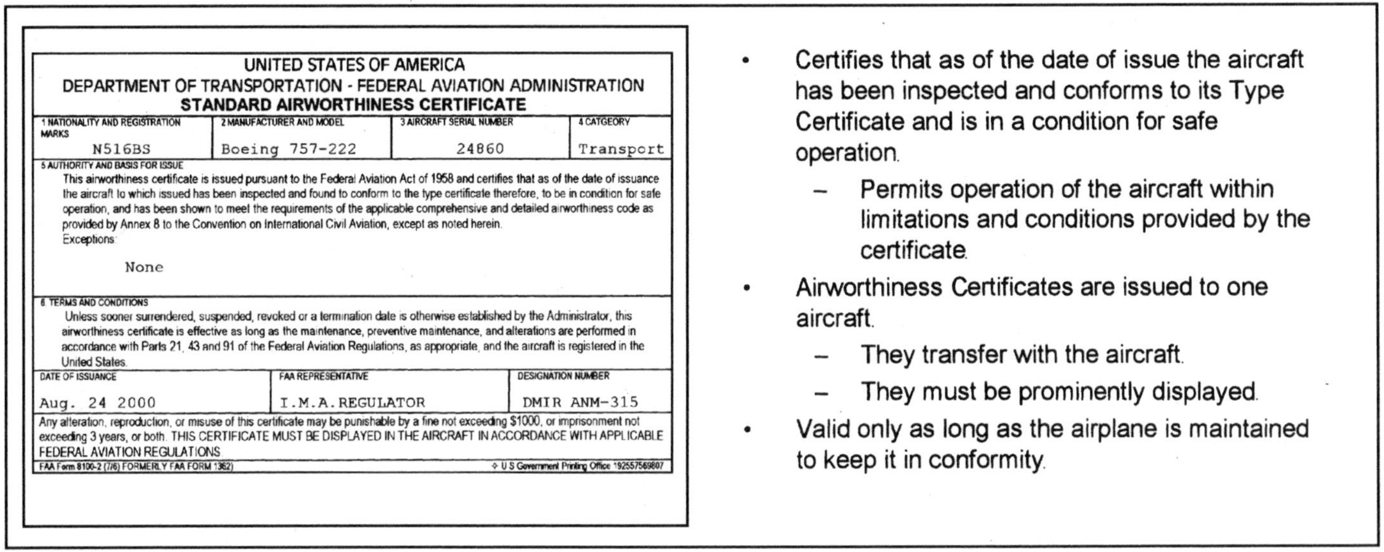

UNITED STATES OF AMERICA
DEPARTMENT OF TRANSPORTATION - FEDERAL AVIATION ADMINISTRATION
STANDARD AIRWORTHINESS CERTIFICATE

1 NATIONALITY AND REGISTRATION MARKS	2 MANUFACTURER AND MODEL	3 AIRCRAFT SERIAL NUMBER	4 CATGEORY
N516BS	Boeing 757-222	24860	Transport

5 AUTHORITY AND BASIS FOR ISSUE
This airworthiness certificate is issued pursuant to the Federal Aviation Act of 1958 and certifies that as of the date of issuance the aircraft to which issued has been inspected and found to conform to the type certificate therefore, to be in condition for safe operation, and has been shown to meet the requirements of the applicable comprehensive and detailed airworthiness code as provided by Annex 8 to the Convention on International Civil Aviation, except as noted herein.
Exceptions:

None

6 TERMS AND CONDITIONS
Unless sooner surrendered, suspended, revoked or a termination date is otherwise established by the Administrator, this airworthiness certificate is effective as long as the maintenance, preventive maintenance, and alterations are performed in accordance with Parts 21, 43 and 91 of the Federal Aviation Regulations, as appropriate, and the aircraft is registered in the United States.

DATE OF ISSUANCE	FAA REPRESENTATIVE	DESIGNATION NUMBER
Aug. 24 2000	I.M.A.REGULATOR	DMIR ANM-315

Any alteration, reproduction, or misuse of this certificate may be punishable by a fine not exceeding $1000, or imprisonment not exceeding 3 years, or both. THIS CERTIFICATE MUST BE DISPLAYED IN THE AIRCRAFT IN ACCORDANCE WITH APPLICABLE FEDERAL AVIATION REGULATIONS

FAA Form 8100-2 (7/6) FORMERLY FAA FORM 1362) ◊ U S Government Printing Office 192557569807

- Certifies that as of the date of issue the aircraft has been inspected and conforms to its Type Certificate and is in a condition for safe operation.
  - Permits operation of the aircraft within limitations and conditions provided by the certificate.
- Airworthiness Certificates are issued to one aircraft.
  - They transfer with the aircraft.
  - They must be prominently displayed.
- Valid only as long as the airplane is maintained to keep it in conformity.

Figure 5-15 *Standard Airworthiness Certificate.*

UNITED STATES OF AMERICA
DEPARTMENT OF TRANSPORTATION - FEDERAL AVIATION ADMINISTRATION
SPECIAL AIRWORTHINESS CERTIFICATE

A	CATEGORY/DESIGNATION		SPECIAL FLIGHT PERMIT
	PURPOSE		MAINTENANCE
B	MANU-FACTURER	NAME	N/A
		ADDRESS	N/A
C	FLIGHT	FROM	SHAWNEE, OKLAHOMA
		TO	DOWNTOWN AIRPARK, OKLAHOMA CITY, OK
D	N- 42565		SERIAL NO. 182-582672
	BUILDER CESSNA		MODEL C-182L
E	DATE OF ISSUANCE 03-22-95		EXPIRY 04-01-95
	OPERATING LIMITATIONS DATED 03-22-95 ARE A PART OF THIS CERTIFICATE		
	SIGNATURE OF FAA REPRESENTATIVE: IMA FRAUD		DESIGNATION OR OFFICE NO.: SW-FSDO-OKC

Any alteration, reproduction or misuse of this certificate may be punishable by a fine not exceeding $1000 or imprisonment not exceeding 3 years or both. THIS CERTIFICATE MUST BE DISPLAYED IN THE AIRCRAFT IN ACCORDANCE WITH APPLICABLE FEDERAL AVIATION REGULATIONS

FAA FORM 8130-7 (10/82) SEE REVERSE SIDE

- Valid only as long as conditions of special issue are met:
  - Experimental
    - Research & development
    - Demonstrate compliance
    - Crew training
    - Exhibition
    - Air racing
    - Market survey
    - Amateur built
  - Restricted
    - New aircraft manufacture FAR 21.25
    - Import FAR 21.185
    - Other FAR 21.29
    - Limited Type Certificate
    - Provisional Type Certificate

Figure 5-16 *Special Airworthiness Certificate.*

The AFM is not a flight crew–usable document. Rather it is best described as a document written by engineers to engineers to fulfill specifically defined legal and regulatory requirements. Its content is rigidly defined in the regulations. It is divided into four sections and a variable number of appendices.

Section 1. Operating limitations contains the certification limits for the airplane such as structural limiting weights, center of gravity limitations, speed limitations, noise certification, limiting altitudes, and temperatures.

Section 2. Emergency operating procedures addressing such things as fire, engine shutdown, depressurization, and emergency descent.

Section 3. Normal operations that actually address special operating procedures commonly classified as nonnormal or abnormal. Material in Section 3 is not the detailed operating information for the airplane.

Section 4. Performance information for the weights, altitudes, temperatures, wind components, and runway gradients, as applicable, within the operational limits of the airplane.

Appendices. These present variable performance information such as engine intermix configurations. A common appendix is the Configuration Deviation List (CDL).[23]

The weight and balance manual

The weight and center of gravity (CG) limits are furnished in the Airplane Flight Manual. For large airplanes, detailed information is provided in a separate weight and balance control and loading document. It is incorporated by reference into the Airplane Flight Manual. The manual, which may be in more than one part includes:

- Generic dimensional and weight data that are necessary for computing the weight and CG and a weighing report.
- The condition of the airplane and the items that are included in the empty weight.
- Loading instructions necessary to ensure loading of the airplane within the weight and center of gravity limits, and to maintain the loading within these limits in flight.

[23]See Chapter 7.

- The weighing report is divided into the recorded airplane weights report and an equipment listing. The equipment listing is a complete listing of all removable equipment installed in the airplane. It includes the specific location of an item by station location, its weight, and corresponding moment.

If the airplane is certified for more than one center of gravity range, the appropriate limitations, for each range, must be included. The weight and balance manual and weight records must be delivered with the airplane. They transfer with other airplane records whenever the airplane is sold or leased.

Instructions for continuing airworthiness

These instructions must include, among other things, the following:

> *Scheduling information (scheduled maintenance) for each part of the airplane and its engines, auxiliary power units, propellers, accessories, instruments, and equipment that provides the recommended periods at which they should be cleaned, inspected, adjusted, tested, and lubricated, and the degree of inspection, the applicable wear tolerances, and work recommended at these periods. The recommended overhaul periods and necessary references to the Airworthiness Limitations which set forth each mandatory replacement time, structural inspection intervals, and related structural inspection procedures. In addition, they must include an inspection program that includes the frequency and extent of the inspections necessary to provide for the continued airworthiness of the airplane.*[24]

This involves a large collection of manuals of varying titles. Contents of the manuals[25] are shown in Figure 5-17. These instructions are the entering argument and basis for the maintenance program required by 14 CFR 121.

> *Accumulated flight hours, calendar time, number of operating cycles or the number of landings are the generally accepted measurements used when specifying maintenance*

[24] 14 CFR 25.1329.

[25] 14 CFR 25, Appendix H.

25xH.3 Content.

The contents of the manual or manuals must be prepared in the English language. The Instructions for Continued Airworthiness must contain the following manuals or sections, as appropriate, and information:

(a) Airplane maintenance manual or section.

(1) Introduction information that includes an explanation of the airplane's features and data to the extent necessary for maintenance or preventive maintenance.

(2) A description of the airplane and its systems and installations including its engines, propellers, and appliances.

(3) Basic control and operation information describing how the airplane components and systems are controlled and how they operate, including any special procedures and limitations that apply.

(4) Servicing information that covers details regarding servicing points, capacities of tanks, reservoirs, types of fluids to be used, pressures applicable to the various systems, location of access panels for inspection and servicing, locations of lubrication points, lubricants to be used, equipment required for servicing, tow instructions and limitations, mooring, jacking, and leveling information.

(b) Maintenance instructions.

(1) Scheduling information for each part of the airplane and its engines, auxiliary power units, propellers, accessories, instruments, and equipment that provides the recommended periods at which they should be cleaned, inspected, adjusted, tested, and lubricated, and the degree of inspection, the applicable wear tolerances, and work recommended at these periods. However, the applicant may refer to an accessory, instrument, or equipment manufacturer as the source of this information if the applicant shows that the item has an exceptionally high degree of complexity requiring specialized maintenance techniques, test equipment, or expertise. The recommended overhaul periods and necessary cross-references to the Airworthiness Limitations section of the manual must also be included. In addition, the applicant must include an inspection program that includes the frequency and extent of the inspections necessary to provide for the continued airworthiness of the airplane.

(2) Troubleshooting information describing probable malfunctions, how to recognize those malfunctions, and the remedial action for those malfunctions.

(3) Information describing the order and method of removing and replacing products and parts with any necessary precautions to be taken.

(4) Other general procedural instructions including procedures for system testing during ground running, symmetry checks, weighing and determining the center of gravity, lifting and shoring, and storage limitations.

(c) Diagrams of structural access plates and information needed to gain access for inspections when access plates are not provided.

(d) Details for the application of special inspection techniques including radiographic and ultrasonic testing where such processes are specified.

(e) Information needed to apply protective treatments to the structure after inspection.

(f) All data relative to structural fasteners such as identification, discard recommendations, and torque values.

(g) A list of special tools needed.

25xH.4 Airworthiness Limitations section.

The Instructions for Continued Airworthiness must contain a section titled Airworthiness Limitations that is segregated and clearly distinguishable from the rest of the document. This section must set forth each mandatory replacement time, structural inspection interval, and related structural inspection procedure approved under § 25.571. If the Instructions for Continued Airworthiness consist of multiple documents, the section required by this paragraph must be included in the principal manual. This section must contain a legible statement in a prominent location that reads: "The Airworthiness Limitations section is FAA approved and specifies maintenance required under §§ 43.16 and 91.403 of the Federal Aviation Regulations unless an alternative program has been FAA approved."

- Appendix H of FAR 25 defines the contents of the Instructions for Continued Airworthiness required by 25.1529
- This translates into a large group of documents including
  - Maintenance Manual
  - Structural Repair Manual
  - Maintenance Planning Document
  - Component Repair Manual
  - NDT Manual
  - Wiring Diagrams
  - etc.

Figure 5-17 *Instructions for Continuing Airworthiness.*

intervals. The selection of a specific parameter is dictated by the particular operating environment encountered.

AD note compliance records

These records must contain a listing of the AD number, the revision date, the method of compliance, and the date/time next action is required.

The manufacturer through engineering changes incorporates the intent of an AD note into new airplanes. Thus, the record provided by the manufacturer will provide the engineering/manufacturing change that met the intent of the AD. The records are permanent for the life of the airplane. They must transfer with the sale/lease of the airplane.

Airplane and component identification

Identification and markings are controlled by 14 CFR 45. Critical components must be permanently marked with a part number or a serial number. A critical part is one for which a replacement time, inspection interval, or related procedure is specified in the Airworthiness Instructions for Continued Airworthiness.

Serial numbers

A serial number is a unique number assigned to only one airplane. It may be thought of as the aeronautical equivalent of an automobile VIN number. It is also used for unique small assemblies such as valves and pumps. Serial numbers are not sequential and will not necessarily identify a given model or type design or component. The manufacturer assigns them. They remain with the unit throughout its life and will not change.

In the case of airplanes, engines, and propellers, they are the only identity differentiating one unit from another of the same kind. Make sure when taking delivery of an airplane be it new, used, or a lease that you verify the serial number on the airplane data plate against the bill of sale, registration, and maintenance records. I know of a new airplane that was delivered that did not have the data plate installed. It was not discovered until years later. I also know of cases of mistaken identity having serious economic consequences to the owners. Consider the following.

An operator contacted the airplane manufacturer requesting a new data plate for his airplane. The reason? The records delivered with the recent purchase of a used airplane did not match the existing data plate. The carrier was informed that a new plate would not be provided. A little investigation revealed that the original owner sold two used airplanes of exactly the same model to two separate buyers on the same day. One airplane was to be placed by the carrier on his Operating Certificate. The other airplane went to a salver and was to be dismantled.

Apparently, somewhere during the life of the airplanes their registration markings were exchanged. My guess was that they were inadvertently switched when the airplanes were painted during refurbishment. Evidently, what happened was that the two buyers identified their purchases by the registry number rather than the serial number. The consequence was that the serviceable airplane went to the scrap yard and was cut up. It is left to the reader to conclude who owned the junk airplane.

Data plates

A fireproof plate (Figure 5-18) that has the following information must identify aircraft and engines:

- Builder's name
- Model designation
- Builder's serial number
- Type certificate number, if any
- Production certificate number, if any
- For aircraft engines, the established rating

The identification plate must be secured such that it will not likely be defaced or removed during normal service, or lost or destroyed in an accident.

Replacement parts

Replacement or modification parts manufactured under a Parts Manufacturer Approval must be permanently marked or have a tag attached containing the following information:

- The letters FAA-PMA
- The name, trademark, or symbol of the holder of the Parts Manufacturer Approval

Figure 5-18 *Typical Data Plate.*

- The part number
- The name and model designation of each type certificated product on which the part is eligible for installation

Registration

Registration markings[26] are displayed on the outside of the airplane. They are frequently referred to as "tail numbers." They are the equivalent of an automobile license plate consisting of an alpha-

[26] 14 CFR 45 and ICAO Annex 7, Aircraft Nationality and Registration Marks.

numeric series. The leading characters identify the country of registry. (Figure 5-19).

Registration, title, and bill of sale topics are covered in 14 CFR 47. These documents are the equivalent of the same type of documents found in the automotive world (Figure 5-20).

Miscellaneous documents

The following documents are not related to or required for airplane certification, registration, and identification. However, they are all required to conduct operations.

Radio station license

The FCC issues a radio station license to each airplane. In the case of airlines this is normally done with a Fleet License that covers all airplanes in the operator's fleet.

SELCAL identification

Selective calling (SELCAL) relieves the flight crew from continuously monitoring traffic control or company radio frequencies during the conduct of a flight. When a facility wants to communicate to the flight crew, the SELCAL for the airplane is activated alerting the crew with an aural chime in the cockpit. Airplanes equipped with selective calling require a unique identification code.

USPHS certificate

The United States Public Health Service (USPHS) issues a certificate of conformance to the airplane. This document attests that the food handling equipment conforms to USPHS regulations.

Additional Reading

Federal Aviation regulations

14 CFR 21, Certification Procedures for Products and Parts

14 CFR 25, Airworthiness Standards: Transport Category Airplanes

14 CFR 33, Airworthiness Standards: Aircraft Engines

14 CFR 34, Fuel Venting and Exhaust Emission Requirements for Turbine Engine Powered Airplanes

14 CFR 36, Noise Standards: Aircraft Type and Airworthiness Certification

Afghanistan	YA	Republic of the Congo	9Q	Lithuania	LY
Algeria	7T	Denmark	OY	Luxembourg	LX
Angola	D2	Djibouti	J2	Madagascar	5R
Antigua and Barbuda	V2	Dominica	J7	Malawi	7QY
Argentina	LQ	Dominican Republic	HI	Malaysia	9M
Argentina	LV	Ecuador	HC	Maldives	8Q
Armenia	EK	Egypt	SU	Mali	TZ
Australia	VH	El Salvador	YS	Malta	9H
Austria	OE	Equatorial Guinea	3C	Marshall Islands	V7
Azerbaijan	4K	Eritrea	E3	Mauritania	5T
Bahamas	C6	Estonia	ES	Mauritius	3B
Bahrain	A9C	Ethiopia	ET	Mexico	XA
Bangladesh	S2	Yugoslavia	YU	Mexico	XB
Barbados	8P	Fiji	DQ	Mexico	XC
Belarus	EW	Finland	OH	Micronesia	VA
Belgium	00	France	F	Monaco	3A
Belize	V3	Gabon	TR	Mozambique	C9
Benin	TY	Gambia	C5	Myanmar	XY
Bhutan	A5	Georgia	4L	Myanmar	XZ
Bolivia	CP	Germany	D	Namibia	V5
Bosnia and	T9	Ghana	9G	Nauru	C2
Herzegovina		Greece	SX	Nepal	9N
Botswana	A2	Grenada	J3	Netherlands -	PH
Brazil	PP	Guatemala	TG	Kingdom	
	PR	Guinea	3X	Aruba	P4
	PT	Guinea-Bissau	J5	Netherlands Antilles	PJ
	PU	Guyana	8R	New Zealand	ZK
Brunei Dar es Saalam	V8	Haiti	HH		ZL
Bulgaria	LZ	Honduras	HR		ZM
Burkina Faso	XT	Hungary	HA	Nicaragua	YN
Burundi	9U	Iceland	TF	Niger	5U
Cambodia	XU	India	VT	Nigeria	5N
Cameroon	TJ	Indonesia	PK	Norway	LN
Canada	C	Iran	EP	Oman	A40
	CF	Iraq	Yl	Pakistan	AP
Cape Verde	D4	Ireland	El	Panama	HP
Central African	TL	Ireland	EJ	Papua New Guinea	P2
Republic		Israel	4X	Paraguay	ZP
Chad	TT	Italy	I	Peru	OB
Chile	CC	Jamaica	6Y	Philippines	RP
China	B	Japan	JA	Poland	SP
Colombia	HK	Jordan	JY	Portugal	CR
Congo	TN	Kazakhstan	UN	Portugal	CS
Costa Rica	TI	Laos	RDPL	Qatar	A7
Cote d'Ivoire	TU	Latvia	YL	Republic of Korea	HL
Croatia	9A	Lebanon	OD	Republic of Moldova	ER
Cuba	CU	Lesotho	7P	Romania	YR
Cyprus	5B	Liberia	EL	Russian Federation	RA
Czech Republic	OK	Libya	5A	Rwanda	9XR
North Korea	P	Liechtenstein	HB	Saint Kitts and Nevis	V4

Figure 5-19 *Aircraft Nationality Marks*

Saint Lucia	J6	Suriname	PZ	Bermuda	VP-B
Saint Vincent -		Swaziland	3D	Cayman Islands	VP-C
Grenadines	J8	Sweden	SE	Falkland Islands	VP-F
Samoa	5W	Switzerland	HB	Gibraltar	VP-G
San Marino	T7	Syria	YK	Virgin Islands	VP-L
Sao Tomé and Principe	S9	Tajikistan	EY	Montserrat	VP-M
Saudi Arabia	HZ	Tanzania	5H	St. Helena/	
Senegal	6V	Thailand	HS	Ascension	VQ-H
Senegal	6W	Macedonia	Z3	Turks and Caicos	VQ-T
Seychelles	S7	Togo	5V	United States	N
Sierra Leone	9L	Tonga	A3	Uruguay	CX
Singapore	9V	Trinidad and Tobago	9Y	Uzbekistan	UK
Slovakia	OM	Tunisia	TS	Vanuatu	YJ
Slovenia	S5	Turkey	TC	Venezuela	YV
Solomon Islands	H4	Turkmenistan	EZ	Viet Nam	XV
Somalia	6O	Uganda	5X	Yemen	7O
South Africa	ZS	Ukraine	UR	Zambia	9J
Spain	EC	United Arab Emirates	A6	Zimbabwe	Z
Sri Lanka	4R	United Kingdom	G		
Sudan	ST	Anguilla	VP-A		

Figure 5-19 *(Continued) Aircraft Nationality Marks.*

FAA Orders

8000.51, Aircraft Certification Directorates

8100.5, Aircraft Certification Directorate Procedures

8110.4A, Type Certification Process

8110.6, Type Certification—Review Cases

8110.21, Supplemental Type Certificate (STC) Approvals, One Aircraft Only

8110.37C, Designated Engineering Representative (DER) Guidance Handbook

8130.2D, Airworthiness Certification of Aircraft and Related Products

8130.21B, Procedures for Completion and Use of FAA Form 8130-3, Airworthiness Approval Tag

8130.24, Procedures for Termination/Nonrenewable of Aircraft Certification Service Designations and Delegations

8300.10, Airworthiness Inspector's Handbook, Volume 2. Aircraft and Equipment, Chapter 89, Special Flight Permit with Continuing Authorization to Conduct Ferry Flights

FAA Advisory Circulars

AC 120-27C, Aircraft Weight and Balance Control

AC 20-65, U.S. Airworthiness Certificates and Authorizations for Operation of Domestic and Foreign Aircraft

AC 20-110J, Index of Aviation Technical Standard Orders

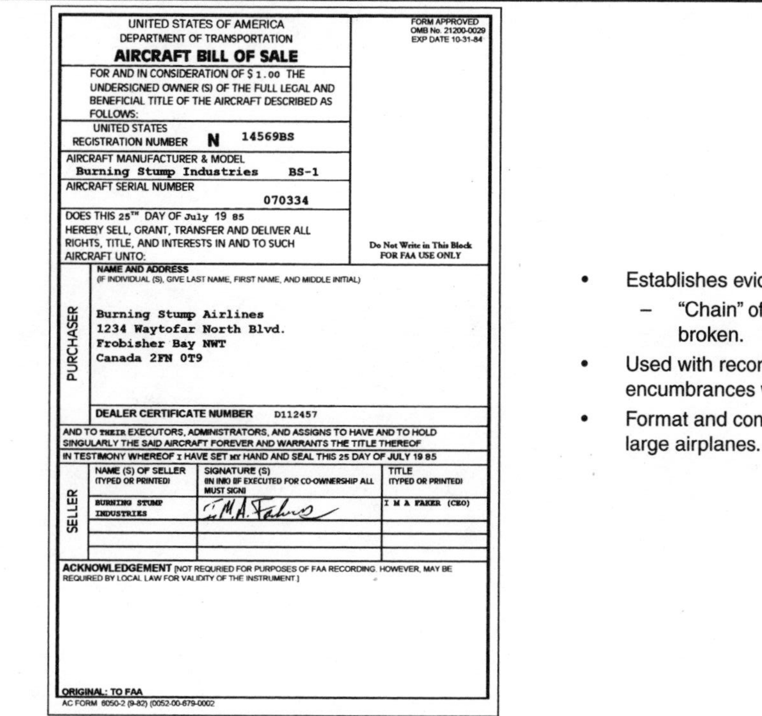

UNITED STATES OF AMERICA
DEPARTMENT OF TRANSPORTATION
AIRCRAFT BILL OF SALE

FORM APPROVED
OMB No. 21200-0029
EXP DATE 10-31-84

FOR AND IN CONSIDERATION OF $ 1.00 THE UNDERSIGNED OWNER (S) OF THE FULL LEGAL AND BENEFICIAL TITLE OF THE AIRCRAFT DESCRIBED AS FOLLOWS:

UNITED STATES REGISTRATION NUMBER **N** 14569BS

AIRCRAFT MANUFACTURER & MODEL
Burning Stump Industries BS-1

AIRCRAFT SERIAL NUMBER
070334

DOES THIS 25TH DAY OF July 19 85
HEREBY SELL, GRANT, TRANSFER AND DELIVER ALL RIGHTS, TITLE, AND INTERESTS IN AND TO SUCH AIRCRAFT UNTO:

Do Not Write in This Block
FOR FAA USE ONLY

PURCHASER

NAME AND ADDRESS
(IF INDIVIDUAL (S), GIVE LAST NAME, FIRST NAME, AND MIDDLE INITIAL)

Burning Stump Airlines
1234 Waytofar North Blvd.
Frobisher Bay NWT
Canada 2FN 0T9

DEALER CERTIFICATE NUMBER D112457

AND TO THEIR EXECUTORS, ADMINISTRATORS, AND ASSIGNS TO HAVE AND TO HOLD SINGULARLY THE SAID AIRCRAFT FOREVER AND WARRANTS THE TITLE THEREOF

IN TESTIMONY WHEREOF I HAVE SET MY HAND AND SEAL THIS 25 DAY OF JULY 19 85

SELLER

NAME (S) OF SELLER (TYPED OR PRINTED)	SIGNATURE (S) (IN INK) (IF EXECUTED FOR CO-OWNERSHIP ALL MUST SIGN)	TITLE (TYPED OR PRINTED)
BURNING STUMP INDUSTRIES	I M A Faker	I M A FAKER (CEO)

ACKNOWLEDGEMENT (NOT REQUIRED FOR PURPOSES OF FAA RECORDING. HOWEVER, MAY BE REQUIRED BY LOCAL LAW FOR VALIDITY OF THE INSTRUMENT.)

ORIGINAL: TO FAA
AC FORM 8050-2 (9-82) (0052-00-679-0002

Figure 5-20 *Aircraft bill of sale.*

AC 21-1B, Production Certificates

AC 21-2H, Export Airworthiness Approval Procedures

AC 21-6A, Production Under Type Certificate Only

AC 21-9A, Manufacturers Reporting Failures, Malfunctions, or Defects

AC 21-12A, Application For U.S. Airworthiness Certificate, FAA Form 8130-6 (OMB 2120-0018)

AC 25.1581-1, Airplane Flight Manual

6

Airworthiness Directives

Introduction

Over the life of aeronautical products, defects affecting airworthiness are discovered. These result from design conditions not foreseen in the original Product Certification or manufacturing deficiencies. Many become evident only after years of in-service operation and are completely unforeseen. Structural fatigue or corrosion are examples.

Any condition that causes the product to be out of conformity must be corrected. The Airworthiness Directive (AD)[1] system is the medium for correction. They are commonly referred to as "AD notes." An AD identifies the disparity, defines its particulars, and establishes limitations for inspection, repair, or alteration under which the product may continue to be operated.

Airworthiness Directives effectively alter the original certification of the product. Thus, for example an AD against a Type Certificated product becomes a part of the Type Certificate.

Threat assessment

Following discovery of a deficiency, the magnitude of its threat to safety is assessed. This may be anything between an immediate hazard and an out-of-conformity condition that in the strict sense causes the product to be unairworthy, but poses no substantive threat to safety. Those conditions that place the airplane or its occupants in

[1]ADs are authorized under 14 CFR 39 and issued under the rulemaking procedures of the Administrative Procedure Act, 5 USC 553, and FAA rulemaking procedures in 14 CFR 11.

peril, require immediate correction. Most commonly, the threat is not grave.

Initial action is to usually call for an inspection along with interim instructions for correction. Detailed corrective actions follow shortly after the results of the inspection are evaluated.

Concurrent with the assessment the safety implications are actions to define the extent of the threat. Does the condition affect all airplanes within a type design or just certain individual airplanes? Are all components affected or just those within a defined group? How are the devices to be identified—part number, serial number, model designation, material type, and so on? This may be determined from manufacturer's records or from the results of an AD that requires an inspection of suspect items. When was the condition discovered? After XXX flight hours, cycles? The problem must have bounds.

Corrective action

Once the problem is clearly defined and the root cause[2] determined, a corrective fix is designed. This involves the FAA, the manufacturer, and selected operators. Factors considered when designing the fix involve more than just the technical design. Additional details requiring attention include:

- The availability of replacement parts must be determined. This affects the effective date and final accomplishment date of the AD. Such things as materials availability and manufacturing lead time must be considered. If immediate or near-term accomplishment is required, then an allocation of available materials must be determined for the affected fleet.
- The methods of accomplishment must be resolved. Type of design and production changes precipitated by the problem may not be readily incorporated into in-service airplanes. Designs must avoid the use of factory tooling or highly specialized machining/fabrication or unique inspection techniques.

[2]Emotion has no place in designing ADs. Unfortunately, the greatest pitfall associated with initiation of an AD is an all too frequent rush to judgment. The temptation to "do something" can be quite strong. It arises from unwarranted political, press, and public pressure before a true cause of the problem is identified. There is ample historical evidence of ADs being issued that did not fix the problem. In some instances, the original condition was exacerbated.

- Accomplishment instructions should avoid detailed disassembly or disturbance of nonaffected components or systems. If not, the probability of consequential damage and maintenance error is increased. The instructions should not introduce the probability of additional failures.
- Airworthiness Directives in which there is no terminating action are particularly onerous. Corrective designs must not include repetitive accomplishment. This may be done in the interim or offered as an alternative, but final terminating action must be designed as quickly as possible.

Repetitive inspections over the life of the airplane increase the following:

- Maintenance cost.
- The probability of maintenance error in performance of the accomplishment instructions.
- The likelihood of inadvertent nonaccomplishment with subsequent enforcement action.

Factors involved in compliance include:

- The means of compliance must possess a degree of flexibility that recognizes varying skill levels, tooling availability, etc. A good fix must allow for alternative or equivalent means of compliance if possible.
- The ability to move the airplane to a location where the required task may be accomplished. If special tooling, inspection, or skills are needed the airplane will more than likely need to be ferried to a station where the requisite facilities are present.
- Compliance time must be balanced against the severity of the threat or airplanes will be grounded unnecessarily.

The AD process

The process is shown in Figure 6-1. When the threat posed is judged minimal, the AD is processed through the normal procedures for rule adoption. A notice of proposed rule making (NPRM) is published in the *Federal Register*. The proposal may be changed or withdrawn on the basis of comments received. The final rule, resulting from the NPRM and the attendant comments, is published as an amendment to 14 CFR 39.

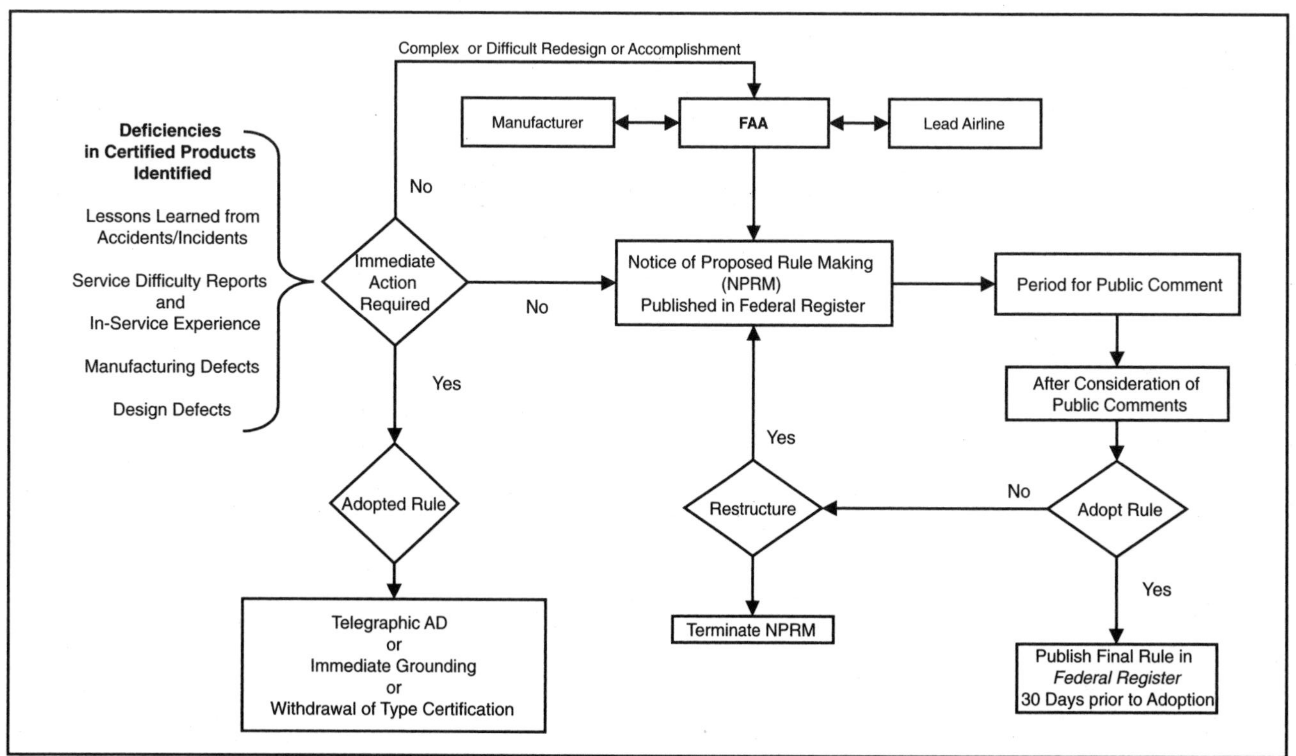

Figure 6-1 *AD Note Process.*

Adopted rules

If the threat posed is urgent, an AD may be adopted without prior notice (without an NPRM). This is done under the emergency procedures granted to the Administrator. It is known as an immediate adopted rule. Each adopted rule AD is normally followed by a final rule version that reflects the final status, changes in accomplishment instruction, product effectivity, and the effective date for accomplishment.

Lead airlines

By the 1990s, operators increasingly complained that the manufacturer's data issued in support of an AD were frequently inaccurate and/or impractical. They further complained about the apparent precipitous actions of the FAA when issuing some ADs. This they thought resulted in rework and/or frequent petitions to the FAA to approve alternative means of accomplishment. Both were costly. In response to these concerns the Air Transport Association (ATA) sponsored a study that evaluated the AD note process. It involved a cooperative effort between the airlines, the FAA, and manufacturers. It included both domestic and foreign operators.

One of the significant findings of the project was that operators believed they did not have enough time or technical data to make a meaningful input to the manufacturers and the FAA before the issuance of an AD. This lack of communication and coordination was thought to be a leading cause of the problems just discussed.

It was decided that deliberation between the parties should begin whenever a potential AD note was identified, long before the FAA became committed to a rulemaking. This, it was believed, would permit a better identification of the problem and evaluation of the issues, thus providing a more workable means of resolution whether an AD was adopted or not.

A significant part of the process involves communication of the manufacturers with the ATA[3] as soon as a potential safety problem is recognized, not after the FAA identifies a need for regulatory action. The is commonly known as the "lead airline" subprocess.

[3]Today rather than just communicating with ATA, manufacturers communicate with non-ATA members as well. This permits the gathering of worldwide input toward the resolution of potential airworthiness problems.

When the process is invoked, one airline is appointed to represent all operators on a specific potential airworthiness problem. This is usually the airline with the greatest interest in the problem.

Distribution of Airworthiness Directives

Adopted rule ADs are distributed "telegraphically" to the known registered owners and operators. Additionally, notification is provided to special interest groups (ATA, International Air Transport Association (IATA), etc.), other government agencies, and civil aviation authorities of foreign countries. Normal Notice of Proposed Rulemaking (NPRM) processed rules are distributed by routine mailing using an FAA-maintained list.

ATA, IATA, Boeing, and Airbus distribute ADs to their customers through their own communications networks. Several commercial regulation publishing services also distribute ADs on a subscription service.

Affected airplanes may be in foreign registry. Registered owners are frequently leasing companies or financial institutions[4] not operators. Airlines frequently lease equipment to each other. Thus, it is difficult to determine the current operator of a product. So despite this extensive distribution system for the documents, they frequently don't get to the right people. It is best not to rely on telegraphic or mail distribution of ADs from all of these agencies. The simplest solution is to review daily the contents of the *Federal Register* for NPRMs and adopted rules. Not receiving an AD through the distribution does not absolve the owner/operator from the responsibility for airworthiness.

The *Summary of Airworthiness Directives* contains all published AD notes from 1940 onward.[5] They are published in six books, three volumes covering small airplanes and three covering large airplanes. Subscription to the summary includes a biweekly supplemental updating service. The summaries are particularly valuable when acquiring a used airplane. Their use permits a full review of outstanding ADs against a given type of design.[6]

[4]It is important that the registered owner(s) of leased aircraft make the AD information available to the operators leasing their aircraft.

[5]See AC 39-6S, Announcement of Availability Summary of Airworthiness Directives details for obtaining these documents. They are available in paper, microfiche, and CD formats.

[6]When reviewing AD notes against a given airplane type make sure that ADs applicable to installed equipment are included. It is very easy to miss some items having outstanding ADs, for example, Time Since Overhaul (TSO) items may not be identified with a specific airplane type.

Understanding the contents of an AD

A typical Airworthiness Directive is shown in Figure 6-2.

Applicability of ADs

An AD contains an applicability statement specifying the product to which it applies. It applies to the make and model set forth in this statement, regardless of the classification of the product or category of the Airworthiness Certificate issued for an aircraft. It applies to each product identified in the statement, regardless of whether it has been modified, altered, or repaired in the subject areas. The presence of any alteration or repair does not remove the product from applicability.

Type Certificate and Airworthiness Certification information is used to identify the product affected. Applicability may be defined by specifying serial numbers or manufacturer's line numbers, part numbers, or other identification. When there is no reference to serial numbers, all serial numbers are affected.

Effective dates

The effective date of the AD or an amendment is be found in the last sentence of the body of each AD. For example, "This amendment becomes effective on July 10, 1995." Similarly, the revision date for an emergency AD distributed by telegram or priority mail is the date it was issued. For example, "Priority Letter AD 95-11-09, issued May 25, 1995, becomes effective upon receipt." The "clock" tracking compliance starts with the effective date.

Compliance time

Compliance with an AD is mandatory. ADs are regulations issued under 14 CFR 39. No person may operate[7] a product to which an AD applies, except in accordance with the conditions of the AD. An airplane that has not had an effective AD accomplished within specified limits is out of conformity and is thus not airworthy. This is consistent with the definition of airworthiness discussed in Chapter 4.

[7]To "operate" is taken to mean not only piloting the aircraft, but also causing or authorizing the product to be used for the purpose of air navigation, with or without the right of legal control as owner, lessee, or otherwise.

96-14-06 Boeing: Amendment 39-9688. Docket 96-NM-134-AD.

{New-96-8}

Applicability: Model 777-200 series airplanes; line positions 1, 3, 5, 7, 8, 9, 11, 12, 13, 15, 16, 17, 19, 20, 21, 22, and 23; certificated in any category.

Note 1: This AD applies to each airplane identified in the preceding applicability provision, regardless of whether it has been otherwise modified, altered, or repaired in the area subject to the requirements of this AD. For airplanes that have been modified, altered, or repaired so that the performance of the requirements of this AD is affected, the owner/operator must request approval for an alternative method of compliance in accordance with paragraph (b) of this AD. The request should include an assessment of the effect of the modification, alteration, or repair on the unsafe condition addressed by this AD; and, if the unsafe condition has not been eliminated, the request should include specific proposed actions to address it.

Compliance: Required as indicated, unless accomplished previously.

To prevent reduced controllability of the airplane due to failure of the lockwire on the bearing retainer nut of the pivot fittings of the horizontal stabilizer, loosening of the retainer nut for the pivot bearing, and subsequent migration of the pivot bearing, accomplish the following:

(a) Within 150 flight cycles after the effective date of this AD: Perform a visual inspection for broken lockwires on the bearing retainer nut of the pivot fittings of the horizontal stabilizer (left and right sides), in accordance with the Accomplishment Instructions of Boeing Alert Service Bulletin 777-55A0003, Revision 1, dated June 20, 1996.

(1) If no broken lockwire is found: Repeat the inspection within 500 flight cycles following accomplishment of the initial inspection. Within 1,000 flight cycles after accomplishment of the initial inspection, modify the bearing nut retention means in accordance with Figure 3 of the alert service bulletin. Following accomplishment of the modification, no further action is required by paragraph (a) of this AD.

(2) If only one broken lockwire is found: Repeat the inspection thereafter at intervals not to exceed 150 flight cycles. Within 450 flight cycles after accomplishment of the initial inspection, modify the bearing nut retention means in accordance with Figure 3 of the alert service bulletin. Following accomplishment of the modification, no further action is required by paragraph (a) of this AD.

(3) If two broken lockwires are found: Repeat the inspection and ensure that the bearing retainer nut is tight thereafter at intervals not to exceed 10 flight cycles. Within 100 flight cycles after accomplishment of the initial inspection, modify the bearing nut retention means in accordance with Figure 3 of the alert service bulletin. Following accomplishment of the modification, no further action is required by paragraph (a) of this AD.

(b) An alternative method of compliance or adjustment of the compliance time that provides an acceptable level of safety may be used if approved by the Manager, Seattle Aircraft Certification Office (ACO), FAA, Transport Airplane Directorate. Operators shall submit their requests through an appropriate FAA Principal Maintenance Inspector, who may add comments and then send it to the Manager, Seattle ACO.

Note 2: Information concerning the existence of approved alternative methods of compliance with this AD, if any, may be obtained from the Seattle ACO.

(c) Special flight permits may be issued in accordance with sections 21.197 and 21.199 of the Federal Aviation Regulations (14 CFR 21.197 and 21.199) to operate the airplane to a location where the requirements of this AD can be accomplished.

Figure 6-2 *Typical Airworthiness Directive*

(d) The actions shall be done in accordance with Boeing Alert Service Bulletin 777-55A0003, Revision 1, dated June 20, 1996. This incorporation by reference was approved by the Director of the Federal Register in accordance with 5 U.S.C. 552(a) and 1 CFR part 51. Copies may be obtained from Boeing Commercial Airplane Group, P.O. Box 3707, Seattle, Washington 98124-2207. Copies may be inspected at the FAA, Transport Airplane Directorate, 1601 Lind Avenue, SW., Renton, Washington; or at the Office of the Federal Register, 800 North Capitol Street, NW., suite 700, Washington, DC.

(e) This amendment becomes effective on July 24, 1996.

Figure 6-2 *(Continued) Typical Airworthiness Directive*

Compliance time is stated in various ways. Typical compliance statements include;

- "Prior to further flight, inspect..."
- "Compliance is required within the next 50 hours time in service after the effective date of this AD..."
- "Within the next 10 landings after the effective date of this AD..."
- "Within 50 cycles..."[8]
- "Within 12 months after the effective date of this AD..."

No person may operate an affected product after expiration of the stated compliance time. In some instances, an AD may authorize operation after the compliance date has passed, if a special flight permit is obtained. These are granted only when the AD specifically permits it. The belief that compliance is only required at the time of a required maintenance inspection, is wrong.

Recurring/periodic ADs

In order to provide for flexibility in administering compliance requirements, an AD should provide for adjustment of repetitive inspection intervals to coincide with inspections required by approved maintenance program inspections. Any conditions and approval requirements under which adjustments may be allowed are stated in the AD. If the AD does not contain such provisions, adjustments are usually not permitted. However, amendment, modification, or adjustment of the terms of the AD may be requested. Exemption procedures are provided in 14 CFR 11.

[8]A cycle refers to a complete operating cycle of an item. It is normally defined an engine start, takeoff operation, landing, and engine shutdown. It may be a pressurization cycle or system operating cycle.

Alternative or equivalent means of compliance

Many ADs indicate the acceptability of one or more alternative methods of compliance. Any alternative method of compliance or adjustment of compliance time other than that listed in the AD must be substantiated to and approved by the FAA before it may be used. Normally the office or person authorized to approve an alternative method of compliance is indicated in the AD.

Consider the following example. A number of years ago it was discovered that it was possible on a particular airplane to inadvertently cross the pressure and return lines in an area of the hydraulic system. The consequences of this cross-connection were judged by the FAA to justify an AD. The tubing connections in question did not have fluid line identification tape adjacent to them. The FAA mandated through the AD that all airplanes (1,850 at the time) in the type design were affected. The manufacturer issued a service bulletin, which was incorporated into the AD by reference.

Now the accomplishment instructions in the bulletin called for disconnecting the affected fittings in the pressure and return lines. Then fluid line identification heat shrinkable sleeves were to be slid onto the lines and the connection reestablished. I leave it to the reader to calculate the probability of crossed lines and/or leaking fittings, which would have resulted from the method of accomplishment described in the bulletin. Operators immediately petitioned the FAA for an alternative means of compliance. This involved using wraparound identification tape thus avoiding breaking open the tubing connections.

In the case of products that have been modified, altered, or repaired so that performance of the requirements[9] of the AD is affected, the owner/operator must develop an alternative method of compliance and request its approval from the FAA. The alternative method may address either no action, if the current configuration eliminates the unsafe condition, or different actions necessary to address the unsafe condition.

[9]Performance of the requirements of the AD is "affected" if an operator is unable to perform those requirements in the manner described in the AD. In short, both the requirements of the AD can be performed as specified in the AD and the specified results can be achieved, or they cannot.

Reference to Manufacturer's Service Bulletins in ADs

Manufacturer's Service Bulletins are normally not related to airworthiness. Incorporation of them is, therefore, not mandatory. However, when a manufacturer's Service Bulletin is incorporated, by reference, into the Airworthiness Directive accomplishment instructions the bulletin becomes mandatory. Thus any change in the details of the bulletin constitutes alternative means of accomplishment. Changes, therefore, must be approved.

Pilot-performed AD checks

Certain ADs permit pilots to perform checks of some items under defined conditions. ADs allowing this action will include specific instructions regarding recording requirements. If the AD does not include recording requirements for the pilot, 14 CFR 43.9 requires that persons complying with an AD make an entry in the maintenance record.

Records

The owner or operator must maintain a record[10] of AD accomplishment. The person who accomplished the action, the person who returned the aircraft to service, and the status of AD compliance are the items of information required to be kept in those records. An audit of the record should be conducted any time the airplane is leased or purchased. This includes new airplanes.[11]

[10]14 CFR 91.417(a) and (b).

[11]An AD changes the certification of the product. Consequently, any items under construction must be altered to correct the discrepant condition. This usually involves a design and production change that will be incorporated at an identified point in the manufacturing cycle, for example, "Line number 431 and on." Line 430 could well be a new airplane not yet delivered. It must have the production equivalent of the AD or the AD note incorporated before delivery.

7

Deferred Maintenance—MEL and CDL

Minimum Equipment Lists (MEL)

The regulations[1] traditionally specified that all installed aircraft equipment required by the airworthiness and operating regulations must be operative. However, experience indicated that, with varying levels of redundancy designed into airplanes, operation of every system or installed component was not necessary when the remaining operative equipment provided an acceptable level of safety.

Beginning in the mid-1950s, the Civil Aeronautics Administration (CAA)[2] granted, only to air carriers, permission to operate with certain items of equipment inoperative; the original intent was to authorize "revenue operations" to a station where repair or replacement could be accomplished. This provided economic airplane utilization and also offered a more reliable flight schedule to the flying public without compromising flight safety.

Structure and approval of these early Minimum Equipment Lists were principally in the hands of the individual CAA Air Carrier Inspectors. Early MELs were separately negotiated between individual operators and their assigned inspectors. Technical evaluation of items for inclusion into a list was based upon the individual inspector's knowledge, competence, and subjective analysis of a specific airplane type.[3] The

[1]Currently FAR 91.205; FAR 121.153 (a) (2); FAR 121.303.

[2]A predecessor organization of today's FAA.

[3]In fairness to the inspectors of the time, expecting them to have the expertise to conduct a thorough analysis of the consequences of a given piece of equipment being inoperative was asking too much.

result was that the MEL for operator A and for operator B (both using the same model airplane) were frequently different; one airline's inspector being very conservative and disallowing all but the simplest of equipment to be inoperative, the other being far more liberal. Individual operators would claim favoritism when one discovered that their competitor had a less restrictive MEL.

This lack of objective analysis and standardization of MELs resulted in the institutionalization of the process by the mid-1960s. The FAA adopted centralized control and publication of Master Minimum Equipment Lists (MMEL). Separate lists are published for each large airplane type.

Present regulations[4] continue to recognize the original MEL concept. In fact, FAR 125, and 135 operators were included in the concept by the late 1970s. 1991 added single engine operations under 14 CFR 135 to the concept. Last, 14 CFR Part 91 operations are also now covered.

Intent of Minimum Equipment Lists

A MEL provides the means to release an airplane for flight with inoperative equipment. The intent is to permit operation for a limited period until repair or replacement of the defective equipment can be accomplished. It is, however, important that repair be accomplished at the earliest opportunity rather than continue operations indefinitely with inoperative equipment.

Nothing in the concept disallows the authority of the pilot in command. The pilot may, at his or her discretion, require that any item covered by the Minimum Equipment List be repaired before flight.

MEL versus the Type Certificate

Operations with certain items of equipment inoperative are not considered an abrogation of the airplane Type Certificate. When operating under 14 CFR 121, 125, or 135, an approved Minimum Equipment List is recognized as an approved change to the type design.[5] Consequently, adoption of an MEL item does not require recertification of the type design.

[4]The applicable regulation for the majority of U.S. air carriers is FAR 121.628. For other types of operators see FAR 125.201, FAR 129.14(b), or FAR 135.179. See FAR 91.213 for special requirements when operating under the general operating rules.

[5]i. e. The airplane's basis of certification, FAR 25 or CAR 4b as amended and shown in the airplane Type Data Sheet.

Operating with an approved Minimum Equipment List and a letter of authorization under 14 CFR 91 constitutes a Supplemental Type Certificate for the airplane. As such an MEL approved under 14 CFR 91 is issued against a specific airplane(s), that is, the airplane serial number(s). The airplane(s) will be listed on the cover sheet of the approved MEL.

Effect of MEL on the airworthiness release or return to service

Is a mechanic or inspector in violation of the FAR for releasing an aircraft as airworthy when certain items are inoperative under an approved MEL? No! For two reasons:

1. An airworthiness release[6] requires certification that the work performed be in accordance with the certificate holder's manual. Inasmuch as an approved MEL is a part of the certificate holder's manual a mechanic is relieved of responsibility for the inoperative status of MEL items. The actions taken under the requirements of an approved MEL would "clear" the discrepancy from the aircraft maintenance record and would consequently revalidate the maintenance release.
2. As has been previously stated an approved MEL is considered a change to the type design. Therefore, the altered status of the airplane under the MEL remains an acceptable certified configuration.

A mechanic is not responsible for any contingent maintenance required by the MEL for any previously deferred items unless additional or repetitive maintenance is required. An airplane maintained under 14 CFR Part 91 will be returned to service under the provisions of 14 CFR 43.5 and is unaffected by this dilemma because an approved MEL under 14 CFR 91 is considered a Supplemental Type Certificate.

The meaning of departure

At issue is an understanding of the meaning of the word *departure* when applying the principles of the MEL. Simply stated when has a flight departed? At block out? Upon completion of a dispatch or

[6]FAR 121.709 permits return to service after maintenance has been performed by the proper execution of a "maintenance release."

flight release? At runway entry? Admittedly, this is hair splitting but it is an issue that has been debated for years. The FAA chief counsel has defined departure to mean "the time the aircraft leaves the take-off surface.[7]

Does the MEL apply to discrepancies occurring after pushback, but before takeoff? The answer according to FAA counsel is yes. "...in the event a discrepancy is discovered following push-back, but before take-off, and the MEL procedures for that item require a mechanic's inspection, take-off would be prohibited. Alternatively, when an item found inoperative during taxi will still permit the aircraft to be operated by the PIC[8] in accordance with an approved MEL, a flight may take off."[9]

Flight Operations Evaluation Boards (FOEB)

The FAA for the development, control, and subsequent revision of Master Minimum Equipment Lists (MMEL) establishes Flight Operations Evaluation Boards (FOEB). An FOEB continues throughout the useful life of the airplane. It typically consists of:

- A chair, who is usually the engineering liaison specialist assigned to the airplane Type Certification project
- The FAA flight test project pilot most familiar with the airplane
- An Air Carrier Maintenance Specialist assigned to the Maintenance Review Board for the airplane
- An Air Carrier Operations Specialist
- An Air Carrier Avionics Specialist assigned to the Maintenance Review Board for the airplane

The chair moreover may, when required, request participation in board proceedings of FAA engineering personnel, Airman Certification Inspectors, or other FAA personnel to meet special needs.

There is no set interval for board meetings. Rather, meetings are called as need arises. For newer model airplanes the meetings might be yearly or even less frequently. A more mature design would require fewer meetings, which thus might be several years apart.

[7]FAA letter to Charles Lewis from FAA assistant chief counsel dated April 17, 1997.

[8]PIC means pilot-in-command.

[9]FAA letter to Charles Lewis, loc. cit.

Master Minimum Equipment Lists (MMEL)

The process of initial adoption and revision of the MMEL is shown diagrammatically Figure 7-1. For a new airplane design, the manufacturer submits a preliminary master list to the Flight Operations Evaluation Board (FOEB). The operators and interested public participants, for example, unions, safety groups, foreign regulatory agencies, and so on, use this in the development of a final master after a period of suitable comment.

FOEB adoption of an MMEL is subject to final review and approval by the chief of Air Carrier Division in Washington D.C. Upon receipt of this approval the original (or revision) to the MMEL is released by the cognizant FOEB. The principal criteria used by the FOEB when adopting an MMEL item are that:

1. An acceptable level of safety is assured after considering subsequent failure of the next critical component within a system.
2. Any interrelationships between allowed inoperative items do not compromise safety.

Once adopted, a list is subject to periodic revision. As operating experience is gained, revisions arise from individual operators petitioning the FOEB through their Principal Operations Inspector (POI) for additions, deletions, or clarifications to the list. There is no defined revision cycle.

There is usually a separate list for each large airplane type; for example, the MMEL for the 737 addresses all model variants of that type design. Small airplanes are covered by a generic master list.

Master MMEL content/format

Figure 7-2 is a typical Master Minimum Equipment List page. The following discusses the format and structure of this page.

"System" numbers are based on those defined by the Air Transport Association of America.[10]

"Sequence" numbers are sequential within each ATA system.

Column 1 "Item" means the equipment, system, component, or function listed in this column.

[10]Air Transport Association of America Specification 100, Specification for Manufacturers' Technical Data.

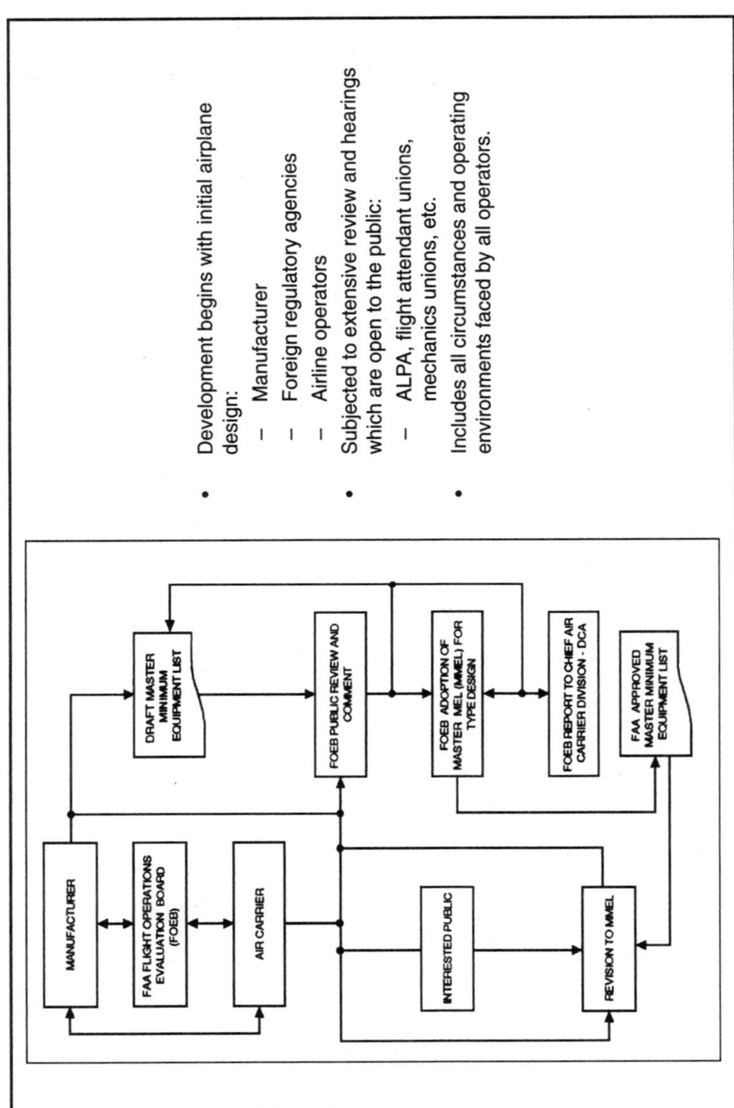

Figure 7-1 *Master MEL development.*

U. S. DEPARTMENT OF TRANSPORTATION FEDERAL AVIATION ADMINISTRATION				MASTER MINIMUM EQUIPMENT LIST
AIRCRAFT: BOEING 737			REVISION NO: 26 DATE: 06/14/89	PAGE: 22-1
SYSTEM & SEQUENCE NUMBERS — ITEM (1.)		2. NUMBER INSTALLED	3. NUMBER REQUIRED FOR DISPATCH	4. REMARKS OR EXCEPTIONS
22 AUTO FLIGHT				
1. Autopilot Systems				
1) ER Operations	C	-	1	*One fully functional autopilot is required for ER operations.
2) Non ER Operations	C	-	0	*May be inoperative unless approach minimums require its use NOTE 1: Any mode which functions normally may be used.
3) Disengage Switches	C	2	1	*(O) One may be inoperative provided the autopilot is not used at less than 1500 feet AGL.
2. Autopilot Disengaged Warning System				
1) Lights	C	2	1	*One may be inoperative when the autopilot is used in any axis.
	C	2	0	*Both may be inoperative if autopilot is not used.
2) Aural Warning	C	1	0	*May be inoperative unless approach minimums require its use.
3. Yaw Damper	C	1	0	*(O) NOTE: Refer to AFM Limitations for SP-77 autopilot.
4. Autothrottle System	C	1	0	*May be inoperative unless approach minimums require its use.

Figure 7-2 *Master MEL content.*

Repair time intervals are defined for all MEL items. Repair of the inoperative system or component must be accomplished within the specified time interval. There are four repair categories:[11] These category letters are placed in the right margin of this column immediately adjacent to column 2.

- *Category A.* Items in this category must be repaired within the interval stated in the operator's approved MEL.
- *Category B.* Items in this category must be repaired within 3 consecutive calendar days (72 hours), excluding the day the malfunction was recorded in the maintenance record (logbook).
- *Category C.* Items in this category must be repaired within 10 consecutive calendar days (240 hours), excluding the day the malfunction was recorded in the maintenance record (logbook).
- *Category D.* Items in this category are those which are in excess of regulatory requirements. They include items that may be installed, deactivated, or removed at the discretion of the operator. They may be added to the operator's MEL but are not required by the MMEL or are required only for a given type of operation. These must be repaired within 120 consecutive calendar days, excluding the day the malfunction was recorded in the maintenance record (logbook).

The * (asterisk) symbol placed in column 4 indicates that the item is not required by regulation but may be installed on some models of the airplane. The item may be installed in the operator's MEL when it has been demonstrated that his aircraft incorporate the device.

Column 2, "Number Installed," is the quantity of items normally installed in the aircraft. It represents the airplane configuration considered when the MMEL item was developed. If the number installed is a variable, for example passenger seats, a number is not entered; rather, the "-" (dash) symbol is entered. When the master list shows a variable number installed, the operator's MEL is required to show the actual number installed in the airplanes.

[11]It is interesting to note that these repair categories are not applicable to an MEL approved under 14 CFR 91.

Column 3, "Number Required for Dispatch," is the minimum quantity of items required for flight operation provided the conditions defined in column 4 are met. When the number required in the master list is variable, i.e."-" symbol is entered in the column, the operator's MEL must show the actual number required for dispatch for the configuration.

Column 4, "Remarks or Exceptions," contains appropriate notes, prohibitions, conditions, and limitations for operation with a specific number of the items, which may be inoperative.

> Notes entered in this column are intended to provide additional information. They identify applicable material, which will assist the user with compliance.
>
> An * (asterisk) symbol placed in this column indicates the item, if inoperative, must be appropriately placarded to inform flight crew members and maintenance personnel of the item's condition.
>
> An (M) symbol in this column indicates that an operator that disarms or controls operation of the inoperative item must develop a specific maintenance procedure. Normally maintenance personnel will accomplish these procedures. However, other personnel may be qualified and authorized to perform certain of the functions. The satisfactory accomplishment of all maintenance procedures, regardless of who performs them, is the responsibility of the operator.
>
> An (O) symbol in this column indicates a requirement for a specific operating procedure that must be accomplished in planning for and/or operating with an inoperative listed item. Normally the flight crew accomplishes these procedures. However, other personnel may be qualified and authorized to perform certain of these procedures. The satisfactory accomplishment of the procedure, without regard to who actually accomplishes it, is the responsibility of the operator.

An entry "As required by FAR" means that the listed item is subject to specific requirements within the operating regulations. The number required by the applicable FAR must be operative for flight operations requiring the item. The list includes:

- Items of equipment or systems related to airworthiness
- Equipment required by the operating regulations
- Other items of equipment that the Administrator finds may be inoperative and yet will permit an acceptable level of

safety to be maintained by appropriate conditions and limitations

The list does not include:

- Obviously required items such as wings, flaps, and rudders
- Obviously not required items such as coffeemakers
- Any deviation from Aircraft Flight Manual Limitations, Emergency Procedures, or Airworthiness Directives

It is important to remember that all equipment related to airworthiness or to the operating regulations not specifically listed in an MMEL must be operative.

Electronic fault alerting systems and the MEL

New generation aircraft display system fault indications to the flight crew by use of computerized display systems. Each aircraft manufacturer incorporates unique design philosophies in determining data to be represented. The "Definition" section of the Master MEL[12] addresses these differences. It provides definitions specific to each manufacturer for use when determining the level of messages that affect the aircraft's release status.

Operator Minimum Equipment Lists (MEL)

MMELs are not intended for operating use. Rather they act as the source document from which an individual operator's MEL is developed (Figure 7-3). FAA procedures[13] require that operators develop their own MEL and obtain approval of it through their Principal Operations Inspector. An individual operator's MEL when appropriately authorized[14] by the administrator permits operation with inoperative equipment for those aircraft listed in his Operations Specifications.

The development of an operator's Minimum Equipment List is shown diagrammatically in Figure 7-3. Operator MELs will frequently differ in format (Figure 7-4) and content from the MMEL but they cannot be

[12]This is contained in definition 23 of that section.

[13]*Air Carrier Inspector's Handbook 8430.6A,* Department of Transportation, Federal Aviation Administration.

[14]Authorization for the use of the MEL is granted by an appropriate entry into Part D of the operator's Operations Specifications.

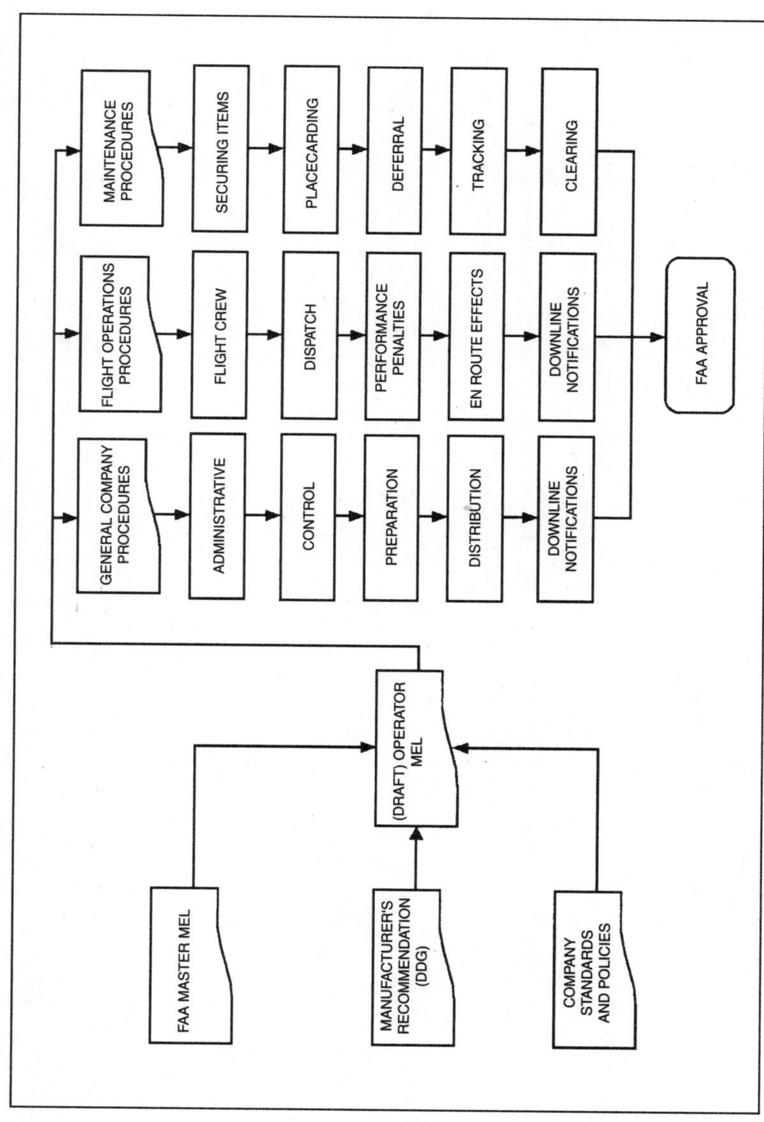

Figure 7-3 *Airline MEL development.*

Burning Stump International Airways
Maintenance Manual
737-2AU Minimum Equipment List

1. SYSTEM & SEQUENCE NUMBERS	ITEM	2. NUMBER INSTALLED	3. NUMBER REQUIRED FOR DISPATCH	4. REMARKS OR EXCEPTIONS
22 AUTO FLIGHT				
1. Autopilot Systems				
1) ER Operations	C	-	1	*One fully functional autopilot is required for ER operations.
2) Non ER Operations	C	-	0	*May be inoperative unless approach minimums require its use
				NOTE 1: Any mode which functions normally may be used.
3) Disengage Switches	C	2	1	*(O) One may be inoperative provided the autopilot is not used at less than 1500 feet AGL.
2. Autopilot Disengaged Warning System				
1) Lights	C	2	1	*One may be inoperative when the autopilot is used in any axis.
	C	2	0	*Both may be inoperative if autopilot is not used.
2) Aural Warning	C	1	0	*May be inoperative unless approach minimums require its use.
3. Yaw Damper	C	1	0	*(O)
				NOTE: Refer to AFM Limitations for SP-77 autopilot.
4. Autothrottle System	C	1	0	*May be inoperative unless approach minimums require its use. See Note 1.

Notes
*Placard and transfer to deferred maintenance log per MM Chapter 8-9-21 and notify MKEBSLM & MKEBSOP per MM Chapter 8-9-23

1. Airplanes N8956BS and N8957BS do not have autothrottles installed.

7/7/89 8-11-9 PAGE 15

- Master MELs are not intended for operational use.
  - Master is a generic document.
- FAA requires each operator to develop his own MEL based upon the MMEL.
- Operator MEL frequently differs in format and content from master.
  - They may contain items not contained in the master. Items that
    - Customize the master to the operator's configuration
    - Add necessary unique operator administrative and maintenance procedures
    - Are not required for a given flight operation
    - Are in excess of FAA requirements
    - Are best placed within the MEL section of the operator's manual, for administrative reasons

Figure 7-4 *Typical Airline MEL page.*

less restrictive. They may include, with appropriate conditions and limitations, items not contained in the master list, such as:

- Equipment not required for a given flight operation
- Equipment that meets more than FAR requirements
- Equipment that, for internal administrative control reasons to the operator, is best placed within the context of the MEL

However, the Administrator through the POI must also approve these nonessential items.

Operators are responsible for exercising the necessary control to ensure that an acceptable level of safety is maintained. This includes a repair program embracing the parts, personnel, facilities, procedures, and schedules to ensure timely clearance of deferred items.

Suitable conditions and limitations in the form of placards, maintenance procedures, crew operating procedures, and other restrictions that are necessary must be specified. In operating with multiple inoperative items, the interrelationships between those items and the effect on airplane operation and crew workload must also be considered.

Review and approval by POI

When reviewing the proposed MEL an operator's assigned POI will check for the following:

- Nothing is contained in the MEL that is less restrictive than the MMEL.
- Nothing contradicts the FAA approved Airplane Flight Manual.
- Nothing violates any special conditions or limitations in any standing Airworthiness Directive (AD) note issued against the airplane.
- Operations (O) and maintenance (M) procedures required by the MMEL are adequate.
- A defined management control process for administration of the MEL that includes:
  - Duties and responsibilities of appropriate management personnel when administering the MEL.
  - Control methods for the document including page control, distribution and revision of it within the operator's system.

- Controls to prevent continued operation with deferred items.
- Administrative procedures to track and account for deferred items within the fleet.
- Procedures that assure repair category time limits will not be exceeded.
- Methods for recording and clearing items in the aircraft maintenance record.
- Placarding scheme used.
- Pilot-in-command notification procedures.
- Inclusion in the manual system of suitable direction and guidance to flight, maintenance, and ground operations personnel addressing use of the MEL.
- Inclusion in both required maintenance and flight operations training curricula of suitable instruction in the use and administration of the MEL.
- Appropriate references to pertinent operating regulations and procedures as well as limitations for the dispatch or the occurrence of a failure en route for continuance of the flight.

When the review is complete and approval is granted, the Principal Operations Inspector endorses the MEL and an appropriate revision is added to Part D of the carrier's Operations Specifications. A typical Operations Specifications' page is shown in Figure 7-5.

The Deferral Process

The system of deferring is shown diagrammatically in Figure 7-5. Specific procedures at any given airline will differ. However, the following is representative of existing procedures.

Once it has been determined that an item is deferrable, a decision is made to defer or fix it. This normally involves, at the minimum, station maintenance personnel and the pilot in command. However, in many instances flight dispatch, maintenance engineering, and a central maintenance control or quality control organization will be a party to the decision. Some airlines designate, in the body of their MEL, specific individuals or organizations that have deferral authority for each item listed in their MEL.

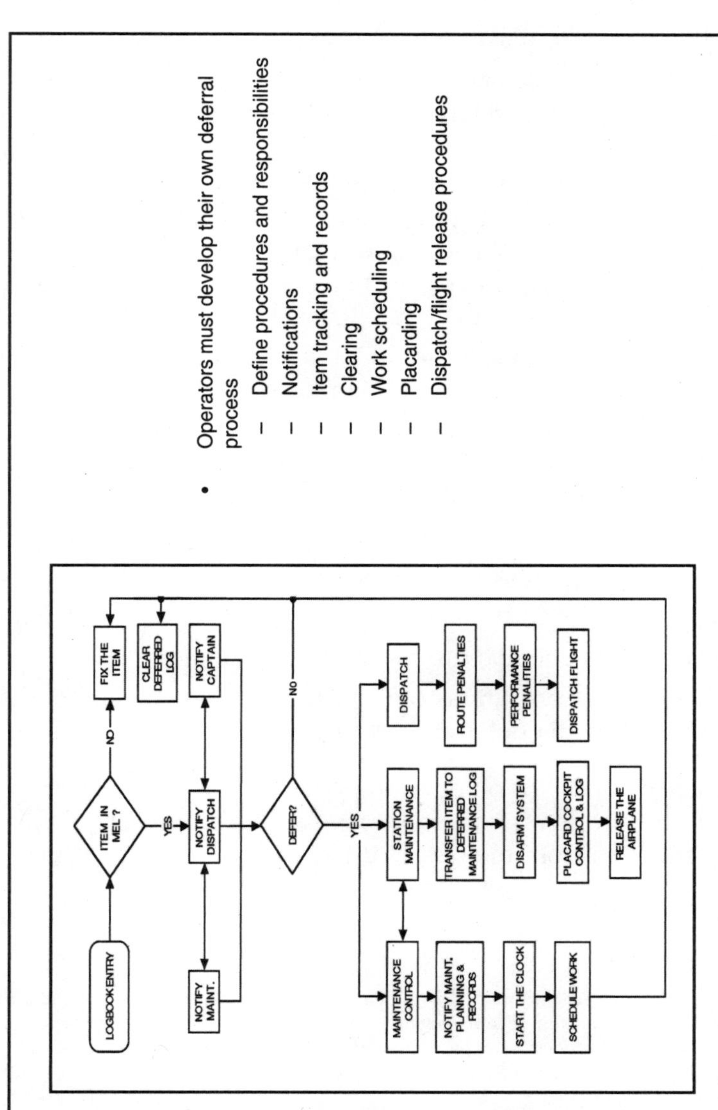

Figure 7-5 *Typical deferral process.*

After the decision is made to defer, specific actions will be taken by various organizations. Station maintenance personnel will:

- Properly secure the deferred item in accordance with company procedures.
- Appropriately, placard the cockpit.
- Clear the aircraft log by transcribing the item from the airplane log to a deferred maintenance log or its equivalent.[15] The deferred log is carried aboard the airplane and is available to the pilot in command.
- Notify the record-keeping function within the airline so that the necessary bookkeeping will take place, thus ensuring that the item is properly tracked and scheduled for later repair within allowable time limits.
- Notify dispatch and/or the pilot in command that the item is deferred.
- Notify any other organizations within maintenance that may be affected by the deferral; for example, main base stores, downline station maintenance, central maintenance control.

Dispatch and/or the pilot in command[16] shall, as appropriate,

- Observe any special limitations or modified operating procedures attendant to the deferred item.
- Notify other operations organizations and downline stations affected by the deferral.

Maintenance control or other appropriate organization charged with tracking deferred items and scheduling will take appropriate action to clear the item from the deferred log within the allowed time for deferral.

[15]The FAA has traditionally required that a deferred item not be carried in the aircraft log but rather that it be transferred from the aircraft record to a separate log. This action returns the aircraft to service "after required maintenance," i.e, revalidates the airworthiness release. Such certification is required prior to operation with any item of equipment that is inoperative.

The use of a separate "Deferred Log" is preferred by flight crews as well as the FAA because it is more convenient to determine the deferred status of the airplane without paging through individual aircraft log entries.

[16]When operating under the supplemental or commercial rules of 14 CFR 121 a dispatch function is not required. Under these circumstances responsibility for appropriate actions rest with the pilot in command.

Configuration Deviation Lists (CDL)

Occasionally nonstructural secondary airframe and engine parts are missing from the exterior of an airplane. This is not as ominous as it sounds. Missing items are normally access doors, small fairings, or nonstructural attachments.

In the past, operators would accomplish simple field repairs. They would cover the openings caused by a missing part either with a sheet metal scab patch or use speed tape[17] and continue operations until a permanent repair could be accomplished. As airplanes became more complex, the impact of missing items on handling qualities came under question. Furthermore, it was thought that in the strictest sense operating an airplane with missing external items violated the type design. The airplane was not airworthy. The FAA disallowed the practice.

But fleet size was increasing, as was daily utilization.[18] The number of instances involving missing parts was consequently increasing. It became a problem of unnecessarily grounding airplanes because of strict interpretation of the regulations. Relief from these circumstances was sought. The argument went something like this. If the item was nonstructural and its presence was for cosmetic or aerodynamic smoothness, and its absence did not threaten the handling qualities of the airplane, then operation should be allowed.

The FAA agreed and institutionalized the identification and control of missing items in the late 1960s. The Configuration Deviation List (CDL) was created. CDL items are contained in an appendix of the approved Airplane Flight Manual (AFM). A typical CDL item list is shown in Figure 7-6.

CDL items versus the Type Certificate

CDL items constitute a certificated configuration and, therefore, should they be missing they need never be repaired or replaced. They are intrinsic to the Type Certificate.

[17]"Speed tape" is the popular name for metal foil adhesive-backed aluminum or stainless steel tape. It is commonly used to externally seal joints or cover holes in structures as well as a host of other applications that are not structural applications. The invention of speed tape in the late 1950s should be considered one of the greatest unrecognized contributions to the aeronautical arts. Its significance to aviation is somewhere between the application of gas turbine engines and the first use of retractable landing gears.

[18]Utilization is the number of hours per day an airplane is flown in revenue service.

Burning Stump Industries
Model 253 Airplane Flight Manual

CERTIFICATE LIMITATIONS

System 23 Communications		1.	Number of Parts Required For All Flight Conditions Except As Provided In Column 2.
ATA SUB-SYSTEM	PART DESCRIPTION EXCEPTIONS		2. REMARKS AND/OR
23-60-	1. Static Dischargers	18	Item 23-60-1 A maximum of 9 static dischargers may be missing with no penalty. At least two dischargers are required on each wing and horizontal stabilizer and one is required on the vertical stabilizer. Where there are only two on a given surface, one of the two must be in the tip position or in the outermost trailing position. NOTE: For airplanes operating where ADF, LORAN or HF communications are not required—all dischargers may be missing with no penalty.
23-60-	1. DME and Marker Beacon Antennae	3	Item 23-61-1 Any number of these items may be missing. No performance penalty.

FAA APPROVED 12-12-93 Appendix CDL Page 5

Figure 7-6 *Typical CDL page.*

Effect of the CDL upon the airworthiness release or return to service

CDL items do not affect the airworthiness release. Return to service occurs only when a missing item is discovered and entered into the log as a discrepancy. Maintenance clearing action in this instance is to invoke the CDL as a valid return to service. Subsequent return to service statements are not necessary.

Qualification of CDL items

Contemporary airplane certification programs include plans for the certification of CDL items. A candidate list of items based on experience with similar designs from other airplanes is developed. It will also include items that exhibit susceptibility to damage or loss. For example, some airplanes include winglets for increased aerodynamic efficiency. These winglets make tempting targets for ground support equipment and other airplanes when in the congested ramp environment of the terminal. Ground damage is quite common.

There must be a full certification testing including a flight test. The airplane must be flown to V_d/M_d (Maximum Dive Speed) with no adverse loads or vibratory conditions shown. Qualification by analysis is rarely allowed.

Adding items to an existing CDL is extremely costly. Moreover, it is frequently difficult to locate an airplane to be used for a flight test after initial airplane certification. Furthermore, the FAA is very conservative and reticent when it comes to adding items without sound technical and economic arguments. There must a compelling economic need, a real problem that has no easy or near-term solution. Thus there must exist an unequivocally demonstrated need before the manufacturer and the FAA will entertain a project to add items.

Handling CDL items

Performance or operating restrictions imposed in the CDL against a given item is a certification limit. They must be observed. When multiple CDL items are carried on an airplane, performance penalties are additive. Managing CDL items should include an administrative process that involves:

- Procedures for notifying dispatch/flight following of the CDL missing parts by an appropriate notation in the aircraft logbook or other means. This should include cockpit placarding similar to that followed with MEL items.
- Appropriate procedures for flight crew.
- Maintenance tracking of the item. Although a CDL item need not be repaired, it is common practice to schedule repair at a

convenient time. Repair removes performance penalties, which over an extended period may have significant economic impact.

Some operators attach a copy of the CDL to their MEL for easy reference. My recommendation is to not do this. MELs and CDLs are separate concepts. Include the CDL in the manual system so that it is easily accessible to both maintenance and flight operations personnel. But they should not be commingled. Doing so in some instances has led to the application of MEL administrative requirements to the CDL. The most undesirable consequence of this is the imposition of time limits on repair.

Dispatch Deviation Guides

As airplanes became more complex, the manufacturer's guidance was increasingly solicited for maintenance and flight operations procedures relating to MEL items. Beginning in the early 1970s manufacturers began to publish this guidance in what is known today as Dispatch Deviation Guides (Figure 7-7).

Many CDL items are difficult to identify. Many are not listed in the manufacturer's Illustrated Parts Catalogs. So CDL item identification was added to the guides. As electronic "glass cockpit" airplanes came into service, Electronic Generated Airplane Fault message lists were added to permit correlation between these messages and deferrable items or systems.

Intent of Dispatch Deviation Guides

The DDG, like the MMEL, is not complete for air carrier use and should not be approved as such. It does not address many of the essential elements required for an operator MEL. Because of this and because responsibility for development of the MEL rests squarely on the operator, DDG information is presented for reference only. It is intended to assist operators when developing the special procedures required by the Master Minimum Equipment List.

There are some that maintain that the only procedures acceptable for an MEL are the manufacturer's. They further maintain that the guides must be FAA approved. That is not the intent of the list. The methods contained in the documents are not presumed to represent

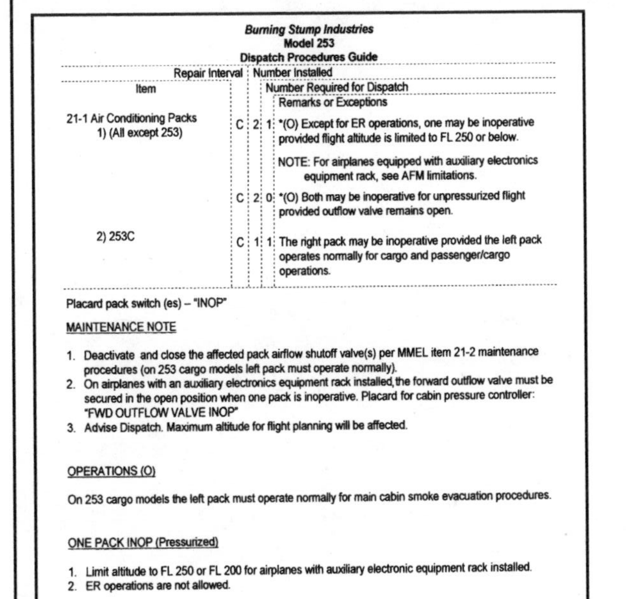

Burning Stump Industries
Model 253
Dispatch Procedures Guide

Item	Repair Interval	Number Installed	Number Required for Dispatch	Remarks or Exceptions
21-1 Air Conditioning Packs 1) (All except 253)	C	2	1	*(O) Except for ER operations, one may be inoperative provided flight altitude is limited to FL 250 or below. NOTE: For airplanes equipped with auxiliary electronics equipment rack, see AFM limitations.
	C	2	0	*(O) Both may be inoperative for unpressurized flight provided outflow valve remains open.
2) 253C	C	1	1	The right pack may be inoperative provided the left pack operates normally for cargo and passenger/cargo operations.

Placard pack switch (es) – "INOP"

MAINTENANCE NOTE

1. Deactivate and close the affected pack airflow shutoff valve(s) per MMEL item 21-2 maintenance procedures (on 253 cargo models left pack must operate normally).
2. On airplanes with an auxiliary electronics equipment rack installed, the forward outflow valve must be secured in the open position when one pack is inoperative. Placard for cabin pressure controller: "FWD OUTFLOW VALVE INOP"
3. Advise Dispatch. Maximum altitude for flight planning will be affected.

OPERATIONS (O)

On 253 cargo models the left pack must operate normally for main cabin smoke evacuation procedures.

ONE PACK INOP (Pressurized)

1. Limit altitude to FL 250 or FL 200 for airplanes with auxiliary electronic equipment rack installed.
2. ER operations are not allowed.

31 May 1990 R-7000/ABG Page 21.1

- Prepared by manufacturers:
  - Contain MEL and CDL reference guidance:
    - Provide manufacturers' suggested flight operations and maintenance procedures associated with MEL
    - Reference material to help identify CDL items
  - Not a working MEL:
    - Intended to assist operator when developing the MEL
    - They are reference documents and will not meet the requirement for an operator MEL

Figure 7-7 *Typical Dispatch Deviation Guide page.*

the only means of accomplishment for flight operations and maintenance to handle a given item.

Operators are able to develop procedures that more closely meet their needs or operating policies. Variables such as airplane configuration, route patterns, special operating procedures, maintenance practices, administrative policies, and so forth, affect a given procedure.

8

Certification Maintenance Requirements

Background

14 CFR 25.1309 adopted in 1970 states in part:

> The airplane systems and associated components, considered separately and in relation to other systems, must be designed so that
>
> (1) The occurrence of any failure condition which would prevent the continued safe flight and landing of the airplane is extremely improbable, and
>
> (2) The occurrence of any other failure conditions which would reduce the capability of the airplane or the ability of the crew to cope with adverse operating conditions is improbable.

This rule introduced into the certification process a structured analysis for identifying hidden failures that would, in combination with one or more other specific failures or events, result in a hazardous or catastrophic failure (Figure 8-1).

In recent years airplane system complexity, integrated digital system architecture, and safety critical functions have increased markedly. Tangled, subtle interrelationships frequently exist. Assessing potential hazards resulting from failures of such systems and their adverse interactions has become harder.

Assessments use both qualitative and quantitative techniques. The predominant method however is quantitative. It is used to support experienced engineering and operational judgment supplementing

Failure Definitions	
Failure	A loss of function, or a malfunction, of a system or a part thereof.
Failure condition	The effect on the airplane and its occupants, both direct and consequential, caused or contributed to by one or more failures, considering relevant adverse operational or environmental conditions. Failure conditions may be classified according to their severity.
Minor failure	Failure conditions that would not significantly reduce airplane safety and that involve crew actions that are well within their capabilities. Minor failure conditions may include, for example, a slight reduction in safety margins or functional capabilities, a slight increase in crew workload, such as routine flight plan changes, or some inconvenience to occupants.
Major failure	Failure conditions that would reduce the capability of the airplane or the ability of the crew to cope with adverse operating conditions to the extent that there would be, for example, a significant reduction in safety margins or functional capabilities, a significant increase in crew workload or in conditions impairing crew efficiency, or discomfort to occupants, possibly including injuries.
Hazardous failure	Failure conditions that would reduce the capability of the airplane or the ability of the crew to cope with adverse operating conditions to the extent that there would be: • A large reduction in safety margins or functional capabilities; • Physical distress or higher workload such that the flightcrew cannot be relied upon to perform their tasks accurately or completely; or • Serious or fatal injury to a relatively small number of the occupants.

Figure 8-1 *Failure definitions.*

Failure Definitions	
Catastrophic failure	Failure conditions that would prevent the continued safe flight and landing of the airplane
Probable failure	Probable failure conditions are those anticipated to occur one or more times during the entire operational life of each airplane. Probable failure conditions are those having a probability on the order of 1×10^{-5} or greater. Minor failure conditions may be probable.
Improbable failure	Improbable failure conditions are divided into two categories as follows: • *Remote failure.* Unlikely to occur to each airplane during its total life but may occur several times when considering the total operational life of a number of airplanes of the same type. Improbable (remote) failure conditions are those having a probability on the order of 1×10^{-5} or less, but greater than on the order of 1×10^{-7}. Major failure conditions must be no more frequent than improbable (remote). • *Extremely remote failure.* Unlikely to occur when considering the total operational life of all airplanes of the same type, but nevertheless has to be considered as being possible. Improbable (extremely remote) failure conditions are those having a probability of on the order of 1×10^{-7} or less, but greater than on the order of 1×10^{-9}. Hazardous failure conditions must be no more frequent than improbable (extremely remote).
Extremely improbable failure	Extremely improbable failure conditions are those so unlikely that they are not anticipated to occur during the entire operational life of all airplanes of one type, and have a probability on the order of 1×10^{-9} or less. Catastrophic failure conditions must be shown to be extremely improbable.

Figure 8-1 *Failure definitions (continued).*

such qualitative analyses. It is largely based on a formal, numerical probability exercise conducted to show design compliance to defined probable failure conditions.

When an analysis shows that required reliability cannot be substantiated, alternatives are available to designers. They may redesign the component or system. This frequently proves to be extremely costly and impractical. The alternative is a reduction of the exposure time to the assumed latent failure. Exposure time to a latent failure is a key element in safety analysis calculations. Limiting the exposure time has a significant effect on the resultant overall failure probability of the component/system. The simplest solution is to impose a special maintenance, time-limited inspection or task that is designed to reveal the hidden failure. This results in failure probabilities that are more acceptable. These periodic inspections are identified as Certification Maintenance Requirements (CMRs). The task intervals are designated as flight hours, cycles, or calendar time, as appropriate.

The underlying goal of any component/system design should be an absolute minimum number of CMRs, with none as the goal. CMRs restart the failure clock to zero for latent failures by verifying that the item has not failed, or cause repair if it has failed. CMRs are not intended to establish supplemental margins of safety for concerns arising late in the type design approval process.

Relationship of a CMR to the Type Certificate

Note well. CMRs are inextricably a part the type design. They are a part of the Type Certificate. They constitute an operating limitation. What are the implications? Simple, if the task is not accomplished at or before the interval required by the certification analysis, the airplane ceases to be airworthy because it is not in conformity to its Basis of Certification. The consequences (with exceptions as shall be seen) to the operator are that the airplane shall not be flown until the required task has been accomplished.

Relationship of CMR to the Maintenance Review Board (MRB) report

A CMR is a required periodic task, established during the certification of the airplane. CMRs are designed to verify that a certain failure has or has not occurred. They do not provide any maintenance function. They are derived from a fundamentally different analysis than that used to define maintenance tasks and intervals that result from Maintenance Steering Group (MSG-3) analysis associated with Maintenance Review Board (MRB) activities.[1]

Maintenance tasks are performed for safety, operational, or economic reasons. CMRs are failure-finding tasks only. They exist to limit the exposure to hidden failures and to fulfill a defined level of reliability.

CMRs versus maintenance programs

CMR tasks are disruptive to orderly maintenance and airplane movement within the flight schedule. They are considered oppressive and costly. Many within the maintenance community believe a CMR is the result of poor design. Historically CMRs have been an evolutionary product of Type Certification. In the past, little if any regard for their implication upon maintenance operations was addressed during design and certification. This led to increasing numbers of CMRs against a given type design. Designs certified in the late 1980s had as many as 33 CMRs attached to the certification. Air carriers became vocal regarding this "CMR explosion." Their dominant objections to CMRs included:

- The inspection intervals and method of accomplishment may not be altered in any way without an amendment to the Type Certificate.
- Inasmuch as CMRs are associated with the Type Certification, they may not be analyzed as a part of a normal MRB[2] process for determining maintenance requirements.
- Inasmuch as the requirement may in no way be altered, no credit may be given for in-service experience. Therefore, normal escalation of the task interval or even possible elimination of the requirement is not allowed.

The consequences to air carrier maintenance programs are unmistakable:

- Increased maintenance administrative, planning, and records cost. What is the cost? One airline stated that CMRs cost $1,000,000 per year per CMR in 1992 dollars.
- CMR tasks are all too frequently out of sequence with the scheduled inspection process derived through the MRB process.

[1]See Chapter 11.

[2]Maintenance Review Board. See Chapter 11.

- Repeated inspections and tests of systems/components "wears them out." Checking them unnecessarily should service experience show, repeatedly, that no faults are detected is an unnecessary burden.
- Revenue flight schedule disruption results. Accomplishment of a CMR frequently results in special routing of the airplane within the schedule pattern in order to accomplish the task at a station having the appropriate mechanics, tooling, inspectors, and so on.
- A worse consequence of lost equipment availability ("lift") arising from "grounding for time" at stations not capable of performing the task. This in turn leads to costly ferry flights to stations where the inspection may be accomplished.
- Equipment substitution for a given assigned flight. If the substitution is a different aircraft type the impact is dreadful because it then affects crew scheduling, passenger services, catering, fleet service, reservations, as well as maintenance,
- In the extreme, flight cancellation or substitute services with another carrier.
- Increased probability of maintenance error arising from repeated inspections.
- Increased exposure to certificate enforcement action should a CMR be inadvertently missed.

Beginning with airplanes certified in the 1990s these problems were addressed through the joint efforts of the manufacturers, FAA, and operators. Airlines were drawn into the design and certification proceedings. A better understanding of the maintenance environment and culture was gained by designers and certification people and vice versa. The result of this collaborative effort has been a removal of the majority of objections to CMRs or at least a workable means of handling them in maintenance programs.[3] CMRs are now classified into two types.

- *One-star CMRs* (*). The tasks and intervals specified are mandatory and cannot be changed, escalated, or deleted without the concurrence of the responsible ACO.

[3]Unfortunately, airplanes certificated before this effort remain burdened with the onerous conditions expressed by the airlines. To remove these obstacles would require an inordinately costly recertification of these airplanes.

- *Two-star CMRs* (**). Task intervals may be adjusted in accordance with an operator's approved escalation practices or an approved reliability program, but the task may not be changed or deleted without prior ACO approval.

Both short- and long-term escalation of CMR inspection time is now possible. Short-term escalation is a temporary extension beyond the required inspection for a specific period. This accommodates uncontrollable or unexpected situations that prevent the CMR from being accomplished within the required interval.[4]

Long-term escalation permits permanent increases in the required inspection time. This allows credit for in-service experience. But instances of a CMR task accomplishment repeatedly finding no failure may not be sufficient justification for deleting the task or increasing the time. This is most likely the case for one-star CMRs.

One-star CMRs are not good candidates for long-term escalation under an air carrier's reliability program. If world fleet service experience indicates that certain assumptions regarding component failure rates made during original certification were overly conservative then one-star CMR tasks or intervals may be changed. Sufficient statistical data to substantiate a change will more than likely not be able to be gathered.

Handling CMRs in maintenance programs

CMRs are listed in a separate document, which is referenced in the Type Certificate Data Sheet. The CMR document is included as an appendix to the MRB report.

Further reading

FAA Advisory Circulars
AC 25.1309-1A, System Design Analysis
AC 120-16C, Continuous Airworthiness Maintenance Programs
AC 120-17B, Maintenance Control by Reliability Methods
AC 121-22, Maintenance Review Board (MRB)
AC 25-19, Certification Maintenance Requirements

[4]Because CMRs are based on statistical averages and reliability rates these extensions do not compromise safety.

9

Aircraft Weight and Balance Control

Introduction

Each operator must have, as a part of its certification, an approved weight and balance control and loading program. The system is usually divided into two functions—a scheme for determining weight and center of gravity (cg) and one for tracking aircraft weight. It includes:

- A load manifest[1] to document loading information and procedures for its preparation
- Procedures giving complete details regarding distribution of passengers, fuel, cargo, and necessary restrictions to passenger movement on the ground and during flight
- Procedures for weighing the airplanes, controlling configuration, and maintaining weight records

For the uninitiated readers the weight of the airplane and its center of gravity is of primary concern. Weight is important in design and operation. In design, certain structural weight limits are defined during airplane certification. In operation, weight defines the required takeoff and landing distances, climb capability, ability to clear obstacles in the flight path, engine out performance, fuel burn, maximum operating altitude. Center of gravity affects the handling qualities of the airplane. Center of gravity of airplanes is

[1]Also known as loading schedule, load sheet, weight manifest.

normally measured and expressed at a percentage of the mean aerodynamic chord (MAC).[2]

Weight and balance system

The loading system is normally the function of the flight operations department at the airline. It consists of a system for computing weight and cg and a means for informing the flight crew of these values before airplane departure. It takes numerous formats. It may involve the flight crew computing the values themselves or sophisticated systems using dispatch/flight following and cargo organizations.

Its form may be a simple trim sheet completed by the crew, small onboard calculators, slide rules, or sophisticated computers relaying the final airplane weight, cg, and takeoff stabilizer trim setting via ACARS[3] downlink to the cockpit.

Maintenance and engineering at the airline is normally responsible for maintaining aircraft weights. This includes actual weighing and the maintenance of records for each airplane.

Weighing requirements

Aircraft weight is normally determined using one of the following procedures:

- *Individual aircraft weight.* The individual weight and cg position of each aircraft is confirmed at specified weighing periods. It is reestablished by computing or reweighing whenever the cumulative change to the operating weight exceeds ±0.5 of 1 percent of the maximum landing weight or the cg varies more than ±0.5 of 1 percent of the MAC.
- *Fleet weight.* The more common form of maintaining aircraft weight in air carrier operations is to use the fleet

[2]Mean aerodynamic chord is the chord of a section of an imaginary airfoil on the wing which would have force vectors throughout the flight range identical to those of the actual wing. MAC is a value used in engineering and weight and balance calculations for convenience. It is used as a reference for locating the relative positions of the wing center of lift and the airplane center of gravity. Ultimately the load distribution determines the static balance and stability of the airplane.

[3]*A*RINC *C*ommunications *A*ddressing and *R*eporting *S*ystem. A system that permits the downlink/uplink of operational data to/from the cockpit. Data are read on a small display unit or may be printed using a cockpit printer.

weight concept. Air carrier airplanes usually have nearly identical configurations for a given airplane model. Therefore a fleet weight, which is an average operating weight, is used.

- Empty fleet weight is usually determined by weighing aircraft according to the sampling allocation shown in Figure 9-1.
- Aircraft weighed are those having the highest time since last weighing. If the operating weight of any aircraft (weighed or calculated) varies more than ±0.5 of 1 percent of the maximum landing weight or the cg varies more than ±0.5 of 1 percent of the MAC from the fleet weight values, it is omitted from the fleet group. It is operated on its actual or calculated weight and cg.

Establishing initial weight

When a new airplane is delivered, the manufacturer provides the weight and balance manual, weighing report, and equipment listing to the purchaser. The operator will adjust these records to account for alterations or modifications after delivery. This determines the operating empty weight (OEW). Common terms are defined in Figure 9-2.

Operators need not weigh aircraft transferred from another operator that has an approved weight and balance program, unless more than 36 calendar months have elapsed since last weighing. Aircraft transferred, purchased, or leased from an operator without an approved weight and balance program can be placed into service without being reweighed if the last weighing was done within the previous 12 calendar months.

Periodic reweighing

The weight of an airplane never decreases over its life. Structural repairs service bulletin modifications and in-house engineering alterations all cause the weight to creep up. Many of these changes are considered negligible and no calculated weight changes are made to the records. However, negligible weight changes are cumulative. With time, their effect upon weight and cg becomes significant. Also,

Fleet Weight Samples	
Fleet Size	**Sample Size**
1 to 3	Weigh all aircraft
4 to 9	Weigh 3 aircraft plus at least 5% of the number over three
>9	Weigh 6 aircraft plus 10% of the number over 9

Figure 9-1 *Fleet weight sampling criteria.*

airplanes collect dirt and debris. This adds to the weight.[4] Airplanes consequently must be reweighed at reasonable intervals.

Aircraft using individual weights are normally reweighed at 36-month intervals. However, the reweighing period may be extended to 48 months. This extension is allowed if the operating weight of any aircraft (weighed or calculated) does not vary more than ±0.5 of 1 percent of the maximum landing weight or the cg varies no more than ±0.5 of 1 percent of the MAC from the previously determined weight and cg.

Aircraft operating under a fleet weight program are reweighed as defined by the program. Because each fleet is normally reestablished every 3 years and a specified sample is weighed at such periods, no additional weighing is considered necessary. A rotation of the aircraft sampled, however, is used so that all aircraft in the fleet will be weighed periodically.

Weight and balance records

The weight and balance system includes methods by which the operator maintains a complete, current, and continuous record of the weight and cg for each aircraft. The records reflect all alterations and changes affecting either the weight or the balance of the aircraft and include a current equipment list. When fleet weights are used, perti-

[4]The weight increase can be large. It consists of internal dirt and debris collecting throughout the structure as well as external dirt. I was once involved in a project which involved weighing a 707 twice. The first weighing was the "as-received airplane." The second weighing was done after the airplane was washed and the interior thoroughly vacuum cleaned. The difference was 300+ pounds.

Selected Weight Terms

Empty weight. The weight of the airframe, engines, propellers, rotors, and fixed equipment.

*Operating Empty Weight (OEW).** The weight established by an operator for a particular model aircraft configuration. It is the empty weight plus standard items such as normal oil quantity, potable water, lavatory fluids, drainable unusable fuel, crew and crew baggage, passenger service equipment such as galley service carts, food, magazines, blankets, headsets. The operator normally defines items included. There may be more than one OEW defined for the airplane. For example, there may be a domestic and overwater OEW. The differences in this example being that the domestic OEW does not include overwater emergency equipment. Specific items for a given OEW are defined in the weight records.

Structural Limits. Weight and cg limits are established at the time of aircraft certification. They are listed in, or referenced by, the type certificate data sheet and in the Airplane Flight Manual (AFM). The operator's weight and balance program should provide for maintaining these limits and should stress the point that the aircraft must be operated at or below its maximum certificated operating weight. The following are general definitions of structural weight limits normally considered in weight and balance programs.

Maximum Zero Fuel Weight. A structural limit. It is the maximum weight of the airplane less fuel and oil.

Maximum Landing Weight. The maximum weight at which the aircraft may be landed. Some aircraft are equipped to jettison fuel to reduce aircraft weight down to the landing limit.

Maximum Takeoff Weight. The maximum allowable weight at the start of the takeoff run.

Maximum Ramp Weight. The maximum allowable weight for taxi.

*Also known as operating weight, or aircraft prepared for service weight.

Figure 9-2 *Selected weight terms.*

nent computations are kept in individual aircraft files. Weight records for an airplane transfer with the records whenever the airplane is sold/leased. They consist of the weight and balance manual for the airplane type, the last weighting report, and the equipment listing.

Standard weights

Operators normally use average weights for passengers, crew, and baggage. Individual operators may derive statistical average weights for use over specific route segments. Typical acceptable average weights are shown in Figure 9-3. Actual weights are used for special passenger groups,[5] cargo, and mail.

Operations Specifications Part E

The Operations Specifications contains the procedures (or refers to the operator's approved weight and balance control program document) used to maintain control of weight and balance of all aircraft operated under the Operating Certificate. This description includes the procedures used for determining weight of passengers, crew, baggage, and cargo; periodic aircraft weighing; type of loading devices; and identification of the aircraft in the program.

[5]Clearly average weight will not work if you are flying a charter group of Summo wrestlers to their annual convention. Other examples of nonstandard groups are athletic teams or symphony orchestras with their instruments.

Typical Average Weights (lbs)					
Military Personnel		**Period**	**Passengers[1]**		
Combat Equipped			**Adult Passenger 180**		**Children[2]**
No	Yes	Summer—May 1	Male	Female	
195	225	through October 31	195	155	80
Flight Crew[3]		Winter—November 1			
Male	Female	through April 30	200	160	80
180	130				

Notes:

1. Includes 20 pounds carry-on baggage for adult passengers.
2. The average weight of children aged 2–12 years normally is used only when needed to accommodate available payload. Otherwise, as passengers, they are considered the same as adults. The weight of children less than 2 years old is factored into adult weight.
3. Or 140 average for all.
4. Operators usually maintain a statistically derived average weight for various route segments or city pairs rather than these stated averages.

Figure 9-3 *Average weights.*

4

Maintenance Practice

10

Maintenance Fundamentals

First principles

Maintenance cannot yield reliability greater than that which is inherent in the design. You cannot improve reliability by trying to inspect it into a design. Inappropriate or inadequate maintenance degrades reliability. Repeated maintenance actions increase wear and the probability of maintenance error with consequent damage. If it ain't broke—don't fix it!

Maintenance rationale

The performance of maintenance restores safety and reliability levels when deterioration or damage has occurred. It does not include servicing.[1] It is performed because the United States Federal Aviation Regulations (FAR) arising from federal law require that:

- "No person may operate an aircraft...unless the mandatory replacement times, inspection intervals and related procedures...have been complied with."[2]
- "The owner or operator, is primarily responsible for maintaining (the) aircraft in an airworthy condition."[3]
- "Each certificate holder (air carrier) is primarily responsible for...the airworthiness of its aircraft...the performance of maintenance...in accordance with its manual and the regulations."[4]

[1]Servicing is the replenishment of consumables needed to keep an item or aircraft in operating condition. It includes such tasks as cleaning, fueling, catering, fluid replenishment, and lavatory service.

[2]14 CFR 91.409(e).

[3]14 CFR 91.403(a).

[4]14 CFR 121.363.

Maintenance does something else. It protects the value of the asset. Large airplanes today cost anywhere from $30,000,000 to well over $100,000,000. Even older airplanes have price tags measured in the millions. Lenders expect the asset to be protected.

Definition of maintenance

As certain items of the airplane structure and its systems deteriorate it is necessary to assure that the design remains airworthy. Maintenance is the action necessary to sustain or restore the integrity and performance of the airplane. It includes inspection, overhaul, repair, preservation, and replacement of parts such as to assure conformity to the Basis of Certification.[5] This is called *continuing airworthiness*.[6]

An admirable definition, but hardly a complete one. It merely says that maintenance organizations mend airplanes. They do more. The maintenance department, at any airline, must also sustain equipment availability, that is, the ability of the airplane to fly the published revenue schedule. The definition of maintenance must include "the management of failure."

Failure

Failure is either the inability to perform within specified limits or the inability to perform an intended function. It may be of a hidden function. Hidden functions are those that are:

- Normally active and whose loss is not evident to the flight crew during performance of normal duties
- Normally inactive and whose readiness to perform, before it is needed, is not evident to the flight crew during performance of their normal duties

The primary consideration of all maintenance decisions is neither the failure of a given item nor the frequency of its occurrence, but rather the consequences of that failure on the airplane and its operation. There are two consequences of failure: those affecting safety and those affecting availability (economics).

[5]See Chapter 4.

[6]*Continuing airworthiness* is a term found throughout the Federal Aviation Regulations. It is frequently, synonymous with the word *maintenance*.

Safety-related failure

Failure that jeopardizes the safety of the airplane or its occupants must be prevented. Flying machines cannot be of such design that any single failure of the device will have catastrophic results.[7] This is aeronautical dogma. Today's airplanes are subject to very few critical failure modes. This safety-related reliability is attributed to the design requirements of the relevant governmental regulations as well as the specifications of operating organizations and manufacturers. Current design practice ensures that vital functions are protected by redundancy, fault tolerance, fail tolerance, and fail-safe features. This assures that, if there is failure, a given function will remain available from other sources to ensure a safe completion of flight.

Economic-related failure

If the loss or deterioration of a particular function endangers neither the equipment nor its occupants, then the consequences of that failure are economic. Examples include systems, components, or features in a design that are not specifically required to demonstrate conformity to the Basis of Certification.

Economic failure should not be a cause for removing equipment from service. Parts experiencing economic failure can and should be scheduled into the normal maintenance routine. This is an idea central to justifying excess features in a type design. If this is not recognized then weight, volume, and cost are added to the design without tangible benefits but with obvious penalty. Finally, consider that this approach promotes not only more economical airplane utilization but offers a more reliable flight schedule to the flying public without compromising flight safety. This is at the core of reliability-centered maintenance.

Failure management

Safety-related failure can be managed. Consider the design that addresses only the avoidance of single catastrophic failures; the airplane and its occupants will not be placed in peril.

But single failures of components or systems will cause the loss of availability of the equipment once the airplane lands. For once a

[7]There are some safety considerations that create an exception to this single failure concept and which require, at least in practice, the accountability for a second failure.

single failure occurs a "no-go" condition arises until repair or replacement is accomplished. There are three design solutions:

1. The components and systems are designed to an exceptional degree of reliability. This is an inordinately costly strategy. Cost-effective design trades must be made between the loss of availability arising from no-go situations and the cost of exceptionally reliable components.
2. If a high degree of reliability is not cost effective, then the design should include a high degree of deferability, that is, a good Minimum Equipment List (MEL).[8] However, this will not totally eliminate no-go conditions.
3. A strategy that assures no-go can be minimized. It involves both a design and a maintenance management technique.

This embraces the incorporation of features that are extra to those required for certification. The predominant strategy for this is the same as that used to avoid safety-related failures; that is, the inclusion of redundancy, fault tolerance, and fail-safe, fail-passive features, but beyond that required to certify the design.

This is not without its price. It increases the number of failure possibilities. It adds more items that can fail. It results in equipment that is more complex and integrated, which makes fault isolation more difficult. It adds to the cost of the airplane.

But this approach, judiciously applied, greatly reduces the consequences of any single failure. Excess features in the design put initial failures of a system into the economic rather than the safety-related failure category.

Fault tolerance

Fault tolerance is one design approach used toward minimizing no-go conditions. It is the built-in capability of a system or Line Replaceable Unit (LRU) to continue to perform its required functions with a specified number of faults. The specified number of faults is dependent on system redundancy and requirements, but at least two or more faults must occur to cause a failure. Fault tolerant design provides uninterrupted and unrestricted operation following random component failures. Fault tolerant systems:

[8] See Chapter 7.

1. Contain at least one spare subassembly or each type in excess of the number required for certification
2. Provide very high fault detection and isolation capability (approaching 100 percent)
3. Automatically reconfigure without any effect noticeable in the flight deck when internal faults are detected to provide continuous, fault-free operation

Preventive maintenance

This is a very misleading term because there are at least three definitions. Unfortunately they are frequently intermixed further adding confusion.

The FAR define preventive maintenance as "...simple or minor preservation operations and the replacement of small standard parts not involving complex assembly operations."[9] This is what is meant throughout the FAR when the phrase is used in the regulations.

Maintenance engineering applies a different meaning to the phrase. The term as used in reliability-centered maintenance is "...the program of scheduled tasks necessary to ensure safe and reliable operation of the equipment. The complete collection of these tasks, together with their assigned intervals, is termed the scheduled maintenance program." Throughout this text I mean the later definition, except when quoting or paraphrasing a specific FAR.

There is a third confusing definition. Preventive maintenance is taken by some when discussing maintenance tasks to mean only those tasks intended to prevent failure. Its use in this context is predominantly associated with the *Hard Time* process discussed in the following paragraphs.

Maintenance processes

Maintenance programs and processes have evolved over the years. There are three recognized processes used in the airline industry to define maintenance: *Hard Time, On Condition,* and *Condition Monitoring.*

[9]14 CFR 1. Examples of preventive maintenance within this definition are contained in 14 CFR 43xA.C.

Hard time

The early generations of air carrier maintenance programs were based on a concept called hard time. This is a life-based concept. It is the oldest maintenance process. It is rooted in the assumption that reliability decreases with increased operating age. Its intended purpose is to prevent failure.

Hard time applies a fixed time and/or cycles[10] that an item is permitted to operate on an airplane. Upon reaching the limit it must be overhauled[11] or replaced (discarded).[12] Items selected for hard time should be limited to:

- Simple items subject to only one failure mode
- Components or assemblies which have definite life limits (for example, metal fatigue) or whose failure could have a direct adverse effect upon safety if they malfunctioned in flight

Hard time maintenance should be avoided:

- It is very costly. By the late 1950s and early 1960s, fully 30 percent of the direct operating cost of the airline was attributable to maintenance. This was in large measure attributable to the dominance of hard time. Compare this to the 10 to 13 percent of contemporary maintenance programs.
- It wears out components overhauling them. Remember the cardinal rule of all maintenance activity.
- It increases the probability of maintenance error causing damage to components or, in the extreme, subsequent catastrophic failures.
- Studies show that a large percentage of the components in an airplane derive no benefit from using hard time. Deterioration and failure are not directly related to time. This

[10]Time relates to operating flight time or in some instances elapsed calendar time. Cycles relate to operating cycles. For example, takeoff and landings or the number of takeoff thrust applications.

[11]Overhaul means restoration of an item to its original service life in accordance with the instructions defined in relevant manuals.

[12]Discard in this sense is removal from service of a device and its subsequent overhaul or repair. It, however, may also mean that the device is removed and scrapped. These include parts and devices for which there are no methods available to adequately assess the serviceability. An example of this is a forging, which may be fatigue-life limited and whose failure can have catastrophic consequences in case of failure. Of course some items are scrapped because it is not economically prudent to repair them.

is particularly true of items that are complex and have multiple failure modes.

- It is extremely difficult to define an initial overhaul time which is well within the service life of the item but which ensures maximum utilization of the available serviceability.

Reliability-centered maintenance (RCM) processes

The general application of the hard time process became outmoded as the industry matured and aircraft became more complex. By the late 1950s, a new methodology evolved that is oriented toward mechanical performance, Reliability-Centered Maintenance (RCM). This is an analytically based concept designed to realize the inherent reliability of a design. It accepts that the operation of a component or system may fail between required inspections, and that the airplane may be safely operated until the next inspection reveals the failure. Its application is, therefore, limited to items whose failure during airplane operation will not have catastrophic consequences. RCM uses two dominant processes.

On condition

On condition avoids predicting hard time failure wear-out points. Rather, repetitive inspections or tests that detect potential failures are adopted These tests call for the removal or repair of individual components "on the condition" that they do not meet a defined standard of performance.

On condition considers specific failure modes. It is based on the likelihood of defining some physical evidence of reduced resistance to the failure mode in question. Until that evidence is present, units remain in service.

A simple example is the coffeemaker. Hard time maintenance would remove it at a specified interval for disassembly, inspection, cleaning, and repair. An on-condition process would specify that the coffeemaker be tested by making coffee.[13] If it does, then the unit continues in service.

[13]Fortunately, maintenance processes don't establish criteria for a quality cup of coffee. Were that the case the electrical shop would be stacked with coffeemakers to the ceiling while the airplanes were equipped with thermos jugs full of coffee from an acceptable outside coffee vendor.

Items and appliances maintained on condition must be restricted to components on which a determination of continued airworthiness may be made by visual inspection, measurements, tests, or other means without a teardown inspection or overhaul. The checks are performed within the time limitations prescribed for the inspection or check. Performance tolerances and wear or deterioration limits are defined in the Instructions for Continuing Airworthiness.

On-condition maintenance can involve bench tests and is thus not restricted to on-airplane inspections. However, on-airplane inspections/tests are preferred.

Condition monitoring

Condition monitoring is either a failure-based maintenance process or a predictive process. The process applies to items that show deterioration over time.

Failure-based monitoring

This is an after-the-fact detection of failures. The process takes two forms under the same name.

In its first form, items are permitted to remain in service without scheduled checks until a functional failure[14] occurs. It may not be used if the failure compromises safety. A simple example defined by extremes demonstrates this. Letting lightbulbs burn out is a perfectly simple application of failure-based monitoring. Letting engine turbine disks go to failure is not. In this case, safety will be compromised and consequential damage can result in extremely costly and lengthy repair.

The second form consists of accomplishing failure-finding tasks to detect hidden failures, for example, engine fire warning tests. Its significant advantage is that maximum economic, safe utilization of equipment is assured. It has two dominant disadvantages:

- Costly consequential damage can occur to other items within the component or system.
- The time of occurrence of the failure is unknown. Consequently, maintenance cannot be planned. Longer out-of-service time may occur because of the unavailability of spares, materials, and people.

[14]Functional failure is defined by how an item failed to perform its function.

Predictive monitoring

Predictive monitoring consists of observing deterioration[15] of a component or system as it trends toward failure. Explicit operating parameters of the device, which are indicative of deterioration or wear, are selected. Collecting and interpreting these data "monitors the condition" of the device.

This form of condition monitoring is best exemplified by engine condition monitoring. Parameters such as altitude, Mach number, inlet pressure and temperature, N_1 and N_2, burner pressure, and EGT are collected. These gas path data are normalized and plotted against time. They are compared against known specific deterioration pattern failure modes evidenced by these parameters. These are coupled with oil sample and vibration analyses.[16] Accurate identification of incipient failures is thus possible, thereby allowing economical repair before the occurrence of extensive costly damage; it is most beneficial with high-cost items such as engine components. Removal, disassembly, or inspection is not required. The principal advantages of predictive monitoring are:

- Incipient failures are identified thereby avoiding loss of airplane availability to the schedule.
- Costly repairs to expensive components can be avoided.
- Better spares inventory control results.
- Improved planning and work allocation is possible.

The principal disadvantage of predictive monitoring is the sizable data collection and analysis requirement imposed. However, this is more than offset by the advantages.

Other maintenance actions

Lubrication

Lubrication may be considered a form of hard time maintenance.

[15]Deterioration itself is a form of failure. Once a device no longer meets the criteria for new design, it has begun the trend toward full failure. It becomes nonairworthy when its wear precludes its ability to perform the requirements of the Type Design, i.e., the device

[16]Oil analysis consists of spectrographic analysis to detect metallic indications of wear. Vibration analysis is used with rotating machinery to detect abnormal mechanical wear or out-of-balance conditions.

Servicing

Servicing is normally accomplished coincident to transit maintenance. Servicing is the replenishment of consumables needed to keep an item or aircraft in operating condition. It includes checking and replenishing oxygen, oil and hydraulic supplies, lightbulbs, tire pressure. These are considered a form of on-condition maintenance.

Replenishment of cabin galleys, the water system, lavatories, the oxygen system, liquor kits, document kits, duty free stores, and fueling, interior cleaning and tidying, etc., are also called servicing. However, they have no maintenance function and are not normally performed by maintenance personnel.

Walk-around inspection

A walk-around inspection is performed at ground level. It is a general inspection intended to identify obvious external structural damage or fluid leaks. Although it is a part of scheduled maintenance, it is not considered a maintenance process. The checks are performed by mechanics before each departure of the airplane from a maintenance station.

There may be independent walk-around inspections completed by the flight crew. These are not considered a part of any scheduled maintenance program.

Zonal inspections

Zonal inspections are not directed at any specific failure mode and thus are not considered a maintenance process. Zonal inspections are intended to assess the general condition, security, and attachment of all items including visible structure in a designated zone, for instance, the cockpit. They are useful to discover accidental damage or missing parts.

Rarely used function inspections

Devices or systems may rarely be used on the airplane. Consequently, evidence of failure will not be readily apparent without occasional operation of the devices. For example, the emergency floor-lighting system. Systems or devices falling within this infrequent-use category must be included in any scheduled maintenance program.

Event-oriented inspections

These are ad hoc special inspections performed after the occurrence of an unusual event. These on-condition inspections assess possible damage after the event. Typical examples are severe turbulence encounters, lightning strikes, bird ingestion or strikes, hard landings, overweight landings, chemical contamination of the airplane by hazardous materials, and engine overtemperature or overspeed. The dominant event unfortunately is ground-support equipment contact with the airplane.

Another form of event-oriented inspection is associated with those initiated by the airline or the manufacturers to identify and/or isolate suspected nonairworthy conditions. Examples include special inspections to locate discrepant fasteners, suspected unapproved parts installations, and missing or incorrectly installed parts.

The cost of maintenance

Maintenance is not free. It accounts for approximately 10 percent of an airline's employees and 10 to 15 percent of its operating expenses. It assures that the airplane is safe and available to fly the revenue schedule by enhancing mechanical schedule reliability thus reducing both the frequency and duration of flight delays and cancellations. Airframe and engine maintenance costs are influenced by many factors as well as accounting and business management decisions. Some of these include the following.

Flight operations

- Operating environments such as marine, desert, and arctic, add special inspections and checks.
- Cultural factors such as type of operations, for example, transporting fish or cattle will increase structural corrosion thereby driving up structural maintenance.
- Average flight length. Longer flights result in high daily utilization but lower cycles. Longer flight length lowers per flight-hour maintenance cost. However, the trip maintenance cost increases with the flight length.

Flight cycle costs are those associated with making the flight regardless of the flight duration. Examples include brake and tire wear.

Flight-hour costs are those associated with the length of the flight. This affects the maintenance cost for systems/components such as avionics, air conditioning, pressurization, equipment, and furnishings.

- *Daily utilization.* Higher utilization increases the frequency of both scheduled and unscheduled maintenance.
- *Schedule and route patterns.* These influence the time available for maintenance, the location, number, and classification of maintenance stations to perform maintenance. This in turn dictates the movement (including transportation cost), location, and quantity of spares, including the transportation cost for spares moving through the system.

Fleet characteristics

- *Airplane age.* Newness reflects the early years of an airplane (usually 3 years). Maintenance costs are lower.

Maturity occurs when the airframe maintenance costs have settled into a constant value and before the effects of aging start. Maturity is considered to start after the completion of the first maintenance cycle. After two or more maintenance cycles the effects of aging due to corrosion control increased structural inspection and increased nonroutine maintenance resulting in higher maintenance costs.

- *Fleet size and composition including commonality of different model types.* Commonality results in fewer part numbers to stock, fewer maintenance plans with their attendant administrative cost, reduced training and learning curves for personnel, and so on.
- *Airplane design quality and maintainability features.* These include a robust Minimum Equipment List, ease of troubleshooting, simple maintenance processes and procedures, and high reliability.

Maintenance practices

- Third-party maintenance vs. in-house. Contract maintenance increases direct maintenance costs. This must be traded against factors such as capital investment in facilities and labor cost to accomplish work in-house

- Maintenance plan—the frequency, number of inspection tasks, and their complexity affect cost. The method of packaging of scheduled maintenance will affect the time out of service. All of the elements of the plan affect the loss of availability (lost revenue) as well as the expenditure of hard maintenance dollars.
- Business image such as extreme cleanliness, passenger appeal, amenities, exterior and interior appearance all contribute to the time out of service, material, and labor costs. Business image extends to a decision regarding acceptable mechanical departure reliability.
- Maintenance philosophy that includes operating policies such as MEL-usage parameters. Some airlines fix everything and allow virtually no deferral.
- Incorporation of airplane and component service bulletins and other modifications.

Accounting practices

- Accounting methods used and economic factors including minimum attractive rate of return, tax and depreciation rates, nonfuel inflation rate, insurance, spares holding factors, maintenance direct labor rate, and burden.
- Cost allocation including such items as common parts and overhead are frequently evenly spread (peanut buttering) over fleet or component type. Real numbers are hard to identify.
- Expense vs. capitalization.
- Warranty credits. How aggressive are warranty items pursued? To what accounts are the proceeds allocated?

Schedule reliability

The maintenance department must assure that the airplanes remain airworthy. But they must also make sure the airplanes once committed to the revenue schedule, remain there without interruption.

Flight disruptions are unacceptable. The most perishable commodity in aviation is an available seat or ton-mile. Lost revenue sometimes referred to as opportunity cost due to cancellation or delays resulting in passenger loss to another carrier are never recoverable. It is directly proportional to the time an airplane is out of service.

Representative delay, cancellation, and out-of-service costs are shown in Figure 10-1. Downline delays and cancellations arising from the first interruption are not included. Admittedly, technical delays and cancellations do not constitute a dominant expense compared to most elements of direct operating cost (DOC).[17] But loss of service is unacceptable. It is poor business practice. The direct economic consequences are major to both the operator and the operator's customer.[18] The cost to the operator includes:

- The loss of revenue
- Poor customer relationships
- Increased spares inventories
- Increased numbers of maintenance stations including skills and personnel required to alleviate such loss
- The costs arising from reroutes, equipment substitution, passenger handling (hotels, buses, meal vouchers, etc.), substitute service, disruption to the airplane plot—an endless list

The cost to the customer takes the form of disrupted plans, missed business appointments, lost time, late shipments, and so forth. People and cargo expect to go to their destination on time.

Typical Single Flight Interruption Cost—US$ (1994)*

Airplane	Delay	Cancellation	Daily Out of Service
Wide body	1000–1500	9,000–10,000	18,000–20,000
Narrow	500–1000	1,500–3,000	5,000–15,000

*May be as high as $160,000–$180,000 for an international flight cancellation

Figure 10-1 *Flight interruption costs.*

[17] Direct operating cost consists of those costs associated with flight crew, fuel, airframe and engine maintenance, insurance, depreciation, and interest.

[18] According to FAA data available for 1994 the calculated annual total operating delay cost to the airlines was $2.5 billion with $7.0 billion attributable losses to passengers. The average delay was 14.23 minutes.

Maintenance disruptions may be divided into two elements: those affected by maintenance operations or those affected by equipment reliability. The responsible agency within the airline is frequently determined by airplane ownership rules.

Schedule reliability, sometimes called dispatch reliability, is the measure of the extent of flight disruptions. It is the number of occurrences a flight is away from the gate on time. It is usually measured as a percentage of the total number of scheduled departures or the total for a given airplane type. Departure reliabilities may also be measured against a given station.

Knowing the cause of a disruption can be an effective management tool. It allows identification of equipment or spares problems, for example. Disruptions are broadly categorized into either noncontrollable or controllable. The former are delays/cancellations due to weather, traffic control, and so on. They are essentially out of the direct control of the operator. The latter relate to the inherent reliability of the airplane and its systems: the effective management of the line maintenance operations or the efficient management of support services such as handling, fueling, catering, crew scheduling, and so forth.

Reliability of equipment is fixed at the time of design. Reliability problems with systems and components will arise. Once reliability becomes an issue because of design deficiencies or low component reliabilities, manufacturers will issue numerous service bulletins against the airplane and components in an attempt to improve it. As shall be seen, these non-safety-related modifications may or may not be incorporated by an operator.

Maintenance operations influence disruptions in many ways. Their influence on schedule reliability may be divided into three general categories. The dividing lines between the three are not sharp. Rather they are blurred and overlapping.

Operating policy and culture

Operating policy is how the operator does, day to day, business in those areas where the FAA has oversight of what is done. The administrative policy and procedures and authority chain for use of the Minimum Equipment List (MEL) varies widely between operators.

What is on time? It is usually accepted in the business that on time means any departure within 15 minutes of the published schedule.

However, this is not a hard rule. Some operators measure on time differently. One airline uses 1 minute rather than 15 as its standard. Some airlines define criteria for short-haul versus long-haul flights. Another uses no tolerance at all. Some don't even measure departure time. Rather they measure on-time arrivals. Arrival reliability actually makes more sense. Passengers and boxes must arrive at a specific time. It really doesn't matter when they depart. A late departure may be irritating to customers but a late arrival is a true disruption to their plans.

Flight disruptions are usually charged against a specific discipline, organization, or station affecting the schedule, for example, weather ATC, catering, crew scheduling, ground handling, maintenance, and passenger services. The categorization of these elements is established by the reporting policies of a specific airline. Some are very detailed, others do not categorize a delay at all. For example, one airline I am familiar with has 86 different categories for flight disruptions; another only accounts for delays by individual station rather than by organization or discipline.

Schedule reliability reports are not always a valid index of what really caused the delay. This is an Achilles heel—accurate reports. There are several factors contributing to accuracy. Interdepartmental rivalries, the "report card" syndrome, possible exposure to regulatory action, pilots versus mechanics versus management conflicts and the thought of being singled out for a delay at the morning meeting blood letting are a few issues that can skew the accuracy of a report. They are all part of the human condition. Organizations and people have personal agendas. For example, many operators have limits for a specific number of deferrals or interruptions per airplane or station within a given period. As these limits are approached, an interesting cultural phenomenon occurs. The number of deferrals and interruptions markedly decline. None of us like to be embarrassed by being a party to a delay. Thus delay reports many times are the product of creative writing and actions rather than raw facts.

I know of a case where, during a delay, the mechanic stole the keys to the catering truck. This gave him a little more time to fix the airplane and the delay was charged to catering, not maintenance. Now the mechanic and the maintenance department didn't get stuck with the delay. Another stunt is to push the airplane back from the gate and then pull it back in. Thus the airplane blocked out on time and then had a gate turnback. The problem must have been with the airplane and not a maintenance delay.

The point of all of this is that it is who calls the flight delay and how the flight delay is called that affects the outcome. Delay reports are very useful tools for isolating schedule problems but they can be skewed and misleading. They must always be read carefully. Sometimes it is not so much what is said in a report as what is not said. Learn how to read between the lines. Don't accept schedule reliability values at face value.

Economics

Economics deals with the cost of doing business. Is 100 percent departure reliability the only acceptable value? Not necessarily. Give me unlimited funds, personnel, and resources and I will give you 100 percent for the factors under my control. This, of course, is not practical. An acceptable schedule reliability figure is a balance between available resources such as equipment, spares, maintenance capability, staffing, and schedule patterns.

Operators establish their own balance between resources expended and schedules reliability. They have an acceptable range for schedule reliability within each of its fleets. Acceptability, many times, is based on the ability to cross-load or download passengers and cargo to other flights and thus not lose revenue. If you are not losing revenue at *X* percent schedule reliability why attempt to achieve a higher value? This is frequently tempered by a business decision regarding public image. Some airlines want 100 percent schedule reliability regardless of the cost.

Even a no cost service bulletin kit can be quite costly, particularly if the airplane must be removed from service or the man-hours for accomplishment are high. Each airline will have criteria for acceptable return on investment (ROI) for determining the non-safety-related service bulletins. Most reliability service bulletins are incorporated by attrition, that is, when the component fails the bulletin is incorporated while it is being repaired.

Management practices

Airline management practices have the greatest influence on schedule reliability. The availability of spares and tooling at a given station, the people available, and turnaround time for maintenance are some of the ways. If you don't have the right tools, the appropriate spares, enough of the right people, and enough time you will have a delay or cancellation.

Airplane utilization

Daily utilization and through-flight ground time directed by marketing management directly effect the time available to maintain the airplane.

Airplane flight patterns

Frequently new routes are opened. Traffic density into a given station will increase before the total support for the operation is in place. The number of days the airplane overnights at a maintenance station influences reliability.

Maintenance station issues

Changing the mix of airplanes at a given station changes mechanic skill, spares, and special tool requirements.

The number of maintenance stations within the system and their individual capability clearly affect timely correction of discrepancies. As schedule patterns and airplane model applications change, the location of designated maintenance stations must also change.

The location or relocation of maintenance stations may be influenced by labor agreements. A new station may be opened in a less-than-desirable location. Many times, senior mechanics will not bid into positions at the new station. This results in less senior and frequently less proficient junior mechanics moving to the new station. I am acquainted with a perfect case, which illustrates this problem. A new model airplane was introduced into service. A group of mechanics underwent training on the new airplane. All were from stations that were to originally receive the airplane into service. Shortly after the start of service, the schedule pattern was changed. Many of the originally trained mechanics did not bid assignment to the new stations. They stayed where they were. The airplane moved. The result was fewer experienced and trained staff available to handle the airplane. Schedule reliability was in steep decline until the new group of mechanics gained experience and training.

Mechanic support

Mechanical departure reliability begs a question, "How bad do you want to go flying?" If you state that you want to meet schedules *X* percent of the time, then how do you get there? Decide who in the maintenance community most influences your ability to get the airplane off the gate on time. Is it the engineering department, the hangars, the shops, stores? No!

The entity that most influences on-time departure is line maintenance—the gate mechanics. They affect repair or deferral. They touch the airplane more frequently than anyone within the maintenance community. They most frequently "return the airplane to service." They work under the most demanding maintenance requirement, that of having the airplane operating within the time restraints of the published revenue schedule. They will give you *X* percent reliability.

The support provided to mechanics (spare parts, special tools, and adequate manuals) affects schedule reliability. The availability and convenience of technical manuals seems like a minor problem, but if the mechanics need information that exceeds the available manuals (including readers or terminals), an airplane must wait.

Maintenance control centers

The use of maintenance control centers definitely helps control interruptions. These centers track problems as airplanes move from station to station and orchestrate the fault isolation. Spares, specialists, and other resources can be propositioned. Without a maintenance control function, each mechanic must start from scratch when the airplane arrives at his or her station unless he or she has contact with the upline station.

Mechanic skill and training

Skilled mechanics, well trained and retrained on specific equipment with specific skills have a significant impact on maintenance delays. Skilled mechanic availability to the marketplace is becoming a very real problem. The vice president of maintenance of one airline recently told me that 25 percent of his mechanics could immediately retire.

If the mix of airplanes and models at a given station is large, it may have been years since a mechanic last saw a currently confronted particular problem.[19] Troubleshooting becomes difficult.

Labor relations

When labor contract renewal time comes around a little "muscle flexing" occurs and schedule interruptions occur. This is most evident when mechanic, pilot, or flight attendant contracts are being

[19]If in fact, the mechanic has ever seen the problem before. One airline made a serious effort to make all the airplanes in the fleet "look the same to the mechanic." They merged with another airline. Shortly after the merger, maintenance delays increased alarmingly.

negotiated. A direct relationship exists between the status of negotiations and the reliability rate.

Spares availability

Today spare parts are extremely expensive. This means provisioning will be conservative. On new models, provisioning is often based on optimistic reliability projections. These two factors frequently result in a shortage of replacement parts.

Schedule reliability associated with maintenance is thus a combination of an acceptable reliability of the airplane coupled with a well-constructed and managed maintenance plan.

Proper delay management has significant impact on airplane availability. Consider the example of an aggressive program to improve maintenance reliability over a 4-year period by one airline. It resulted in the following gains.

The mean time between unit removal (MTBUR) for the airplane went from 8.5 hours to 15.1 hours for 600 components that were tracked. During this same period utilization went from 8 to 9 hours per day (a 12.5 percent increase in productivity). All this translated into the following

- For each airplane in the fleet, 125 fewer parts (37 percent) were being removed while utilization increased 1 hour.
- For an average utilization of 8.5 hours per day, 160 fewer parts were removed from each airplane.

The financial benefits were large. For example, one operator with 20 airplanes would have had 7,300 more flight-hours available at the end of the program while saving maintenance and spares cost by removing 2,500 fewer parts from the airplanes.

Digital airplane maintenance

Built-in test equipment (BITE)

BITE consists of imbedded monitors and software in Line Replaceable Units (LRUs) that assess their condition and evaluate their performance. It

- Permits troubleshooting of complex integrated systems

- Contributes to the use of condition monitoring
- Provides detailed engineering data to designers

Digital airplanes

"Digital" or "glass cockpit" airplanes include in their design:

- Components containing microprocessors which replace many mechanical or electromechanical functions. This gives increased component reliability
- Integrated functions between individual components and systems.
- Digital-based BITE[20] monitors consisting of intricate circuits and logic to assess not only individual LRUs but also the elaborate interrelationships between them.

These monitors constantly check the health of a unit as opposed to mechanic-initiated tests found in previous BITE designs. When a failure appears, the circuitry generates electronic fault messages that are displayed on display screens in the cockpit. These messages tell the flight crew and mechanics of malfunctions occurring within the unit/system.

- Diagnostic information to mechanics through either initiated tests or fault reports automatically generated by BITE monitored devices.
- Integrated BITE-based maintenance systems. These have the ability to effectively monitor the deterioration of features and manage the scheduling of their repair while assuring airplane safety and availability to the revenue schedule.

Limitations of digital BITE

Digital BITE is skillful at isolating failure causes. But it is not infallible. To understand BITE it is necessary to understand some of its characteristic limitations and design deficiencies.

Some are intrinsic, others are design related. It is important to know what BITE-based systems can do. But it is more important to appreciate BITE limitations. It is necessary to grasp these limita-

[20]These functions are fully integrated into the device while isolated from any safety-critical functions.

tions in order to understand why BITE can "lie" and further how these limitations affect the interpretation of BITE data during maintenance.

Digital airplanes introduced during the 1980s were plagued by nuisance messages,[21] confusing intelligence, poor displays, superfluous messages, and too much information. This gave BITE-based systems a bad name. In the intervening years, the utility of BITE has improved. The designs have been refined eliminating many problems through software improvements, better logic, and more accommodating circuit designs.

The maintenance community has a better understanding of the designs and limitations. It the case of earlier designs the improvement, in many instances, has not been design improvement as much as it has been the ability of maintenance groups to work around known problems.

But economics preclude complete redesign and retrofit of earlier systems to completely correct these designs. Thus, these airplanes will never have the reliable BITE that is available in airplanes that entered service in the 1990s. The newer fully digital airplanes have largely eliminated, or at least control, many of the following deficiencies.

Intrinsic limitations

BITE is subject to disturbances and design limitations that will cause it to generate incorrect fault reports (lies). These intrinsic factors may be controlled, but they will never be eliminated.

Environmental effects. Environmental concerns include a whole list of causal factors. For example the conformal coatings used to protect circuit boards is frequently breached. This allows moisture and other contaminants to penetrate the circuitry that can lead to intermittent performance, grounding, and so forth. This can produce fault messages that are not true.

The limitations of scale BITE-based systems indicate only the integrity of that portion of the component/system they monitor. BITE is further limited by:

- The amount of fault logic and the degree of fault consolidation logic included

[21]Nuisance messages are those generated by BITE, which are not indicative of an actual fault.

- The parameters to be monitored (the number and kind of sensors)
- The budget allocated to it during design of a component

Coincidentally, the BITE itself must not become a source of failure. After all, the primary function of any component is to perform the intended design function, not to check itself for failure.

Because of this, BITE is often unable to absolutely identify a single faulty component within a system. The solution can be more sensors with more fault logic. This permits better isolation of discrepant LRUs. But it also leads to more wire, more weight, and more fault messages; this in turn means more chances for logic errors and "lies." There is a finite limit to the amount of sensing and logic. Thus, a limit exists to the fault intelligence that may be derived from any BITE-based system. Consequently, BITE results will always contain some ambiguity.

Fault logic and consolidation Frequently, elusive, subtle relationships exist within and between systems. As an example, consider the condition where a faulted component within a system is unable to provide necessary information to another downstream LRU. If BITE logic does not consider this, the consequence is multiple fault reports: the upstream truly failed component and the downstream units dependent upon the first component's information. This is called a cascaded fault. Cascaded faults confront the mechanic with a dilemma: What do I fix? They can be controlled by design.

Fault logic in any system must effectively consolidate fault reports from many components within a system and eliminate or at least control cascade effects. The number of fault monitors in the design and the consolidation logic again limits how well this is done. If it is done poorly or not at all, the perception is again that BITE "lies."

Ambiguity The inherent limitations of sensor logic and design frequently make it impossible, when consolidating faults, to absolutely identify a single component or system as the root cause of a malfunction. In most instances, the solution is carried to a group of components that are the possible causes. This is known as an ambiguity group.

In many designs ambiguity groups are not given to the mechanic. In these situations, the designer has made a decision for the mechanic on the basis of his knowledge of the system, its components, individual reliability, and his perception of how mechanics troubleshoot. Information is withheld from the mechanic that may be beneficial to his diagnosis of the problem. BITE should never make decisions for the mechanic. The system must either unequivocally isolate the fault or, in the absence of this, tell the mechanic the ambiguity.

Fault correlation BITE monitors and BITE logic in different airplane designs send reports to the flight deck and to the mechanic. There must be a one-to-one correlation between the intelligence provided the flight deck and the message given the mechanic and the underlying root cause to permit isolation of the root causes of a discrepancy. It is completely unacceptable to have a BITE-based system that tells the flight deck something is wrong and does not tell the mechanic what it is. Later designs have better fault correlation.

Hair triggering This is a design problem rather than an inherent characteristic. Hair triggering is the premature setting of a BITE message. It is caused by how a monitor sets a fault. If the fault monitor logic does not "sample" the occurrence of a fault, the logic will send a message in response to transient conditions rather than a true faulted condition. This results in nuisance messages.

Power sensitivity Digital circuits are sensitive to power interrupts, voltage transients, timing delays, electrical bus power-up sequences, and the like (dirty power).

Fault monitors with insufficient time delays and consolidation logic will trigger nuisance faults. They are an annoying cause of delays most evident when the airplane is first powered or when it encounters a power transfer. It is well known that the majority of these faults occur just prior to pushback. Once the airplane is away from the gate, these nuisance faults rarely occur. Contemporary systems are able to handle "dirty" power.

Interface issues

Information Display Unfortunately, the methods of displaying BITE results remain essentially unchanged from the previous generation of airplanes. BITE messages are diverse and nonstandardized in their presentation. Box designers and component vendors interpret in their own way the few display criteria that exist. No common standard of performance exists. Much BITE is still on the front of the box and it still consists of lights and unintelligible codes. This problem is compounded because the number of boxes with BITE capability has increased markedly. Airplanes using central maintenance computers do a much better job of solving this problem.

Intended user As BITE capability grew, many disciplines saw benefit. Consequently designers tried to satisfy everyone's needs: their own design needs, engineering at the airline and the manufacturer, hanger maintenance, bench mechanics and line mechanics, maintenance planners, and statisticians. Now no device can be truly successful if it tries to be all things to all people.

One of the biggest mistakes the industry made in the development of BITE is that it never really answered the question, "Who is the principal user and beneficiary of BITE?" This is central to a successful system. The system may accommodate all users but it must be hierarchical. The principal user should always be defined as the gate mechanic, then engineering at the airline, and finally designers.

Fault message relevance It is now possible to obtain large amounts of data whether the data have any merit or not. Unfortunately, little thought has been given to defining the information needed. It is now possible to present fault messages that convey more than just simple diagnostics. The maintenance community is overwhelmed with information of dubious benefit. The design must be bounded to make sure that the users of the information will derive benefit from, or be confused by, fault messages. The former is the only acceptable design.

Fault message differentiation Many fault reports are not differentiated between those annunciating airworthiness loss and those that are identifying deterioration or economic failure. The system must do this to avoid unnecessary maintenance actions when the airplane is committed to the schedule.

BITE messages annunciating economic failure should clearly be differentiated from safety-related failure. Designs from the 1980s do a poor job of this.

BITE as troubleshooting tool

The concept of digital maintenance has been oversold. BITE-based maintenance systems will not replace skilled mechanics. It will not unequivocally identify failed LRUs except in a few instances.

Fault isolation (troubleshooting) is an art not a science. It requires a diagnostician with an intuitive sense and judgment coupled with a detailed knowledge of

- Systems and their component parts
- The role each plays
- Their distinctive behavior and inherent design limits
- The effect of disturbing influences upon those systems and their component parts

Good troubleshooting is nothing more than good deductive reasoning; at the center of that reasoning is a careful collection and evaluation of physical evidence. Many of today's devices use computer chips to provide a function formerly fulfilled by substantial mechanical parts or subsystems. Consequently troubleshooting—in the traditional sense of searching out physical evidence of failure—is hindered.

Unfortunately, you can't troubleshoot a computer chip by looking for physical evidence of failure. A broken chip doesn't look any different that a healthy one. Although it can be argued that broken chips occasionally make smoke, evidence of malfunction is seldom readily apparent. Broken chips don't leak, vibrate, or make noise. Bad software within them doesn't leave puddles or stains as evidence of its misbehavior. 1s and 0s falling off the end of a connector pin are difficult to see. Before one may look for physical clues, it is necessary to have a means to give "eyes" to the mechanic for the unseen elements within the design. These eyes are BITE-based maintenance systems.

You need a digital device to translate digital information into understandable intelligence for human beings. BITE is thus nothing more than another, now necessary, tool in a mechanic's toolbox. It is not an automatic troubleshooter. BITE cannot resolve ambiguity. It does not have intuitive skills. It does not have knowledge of the idiosyncrasies of designs or the personality of individual airplanes. It doesn't know the routing of the airplane. It doesn't know what was changed yesterday. It cannot think. And above all else, it does not have a licensee allowing it to return to service.

11

Scheduled Maintenance Programs

Introduction

The airworthiness regulations identify a requirement titled "Instructions for Continuing Airworthiness" (FAR 25.1529 and FAR 25, Appendix H). Developing the scheduled maintenance plan for a new airplane is long and very costly. It takes several man years of work, thousands of pages of documentation, and numerous meetings between the participants. The process can take 3 to 5 years to complete. The maintenance program development concurrent with the airplane development and certification is shown in Figure 11-1.

Consider that for the 777 airplane the task of defining tasks was begun in the early 1990s. The original Maintenance Review Board (MRB) document was adopted by the FAA in 1995 just 2 weeks before the airplane certified. The project involved several hundred people from all over the world. It included mechanics, design engineers, maintenance engineers, regulators, and countless other skills. Scheduled maintenance may be divided into several elements:

- Component and system elements involving inspection and check of individual components and complete systems against established performance standards
- Structural elements consisting of a combination of visual inspections and nondestructive test methods at stated intervals
- Corrosion control and inspection
- Lubrication elements defining intervals for lubrication of various installed components

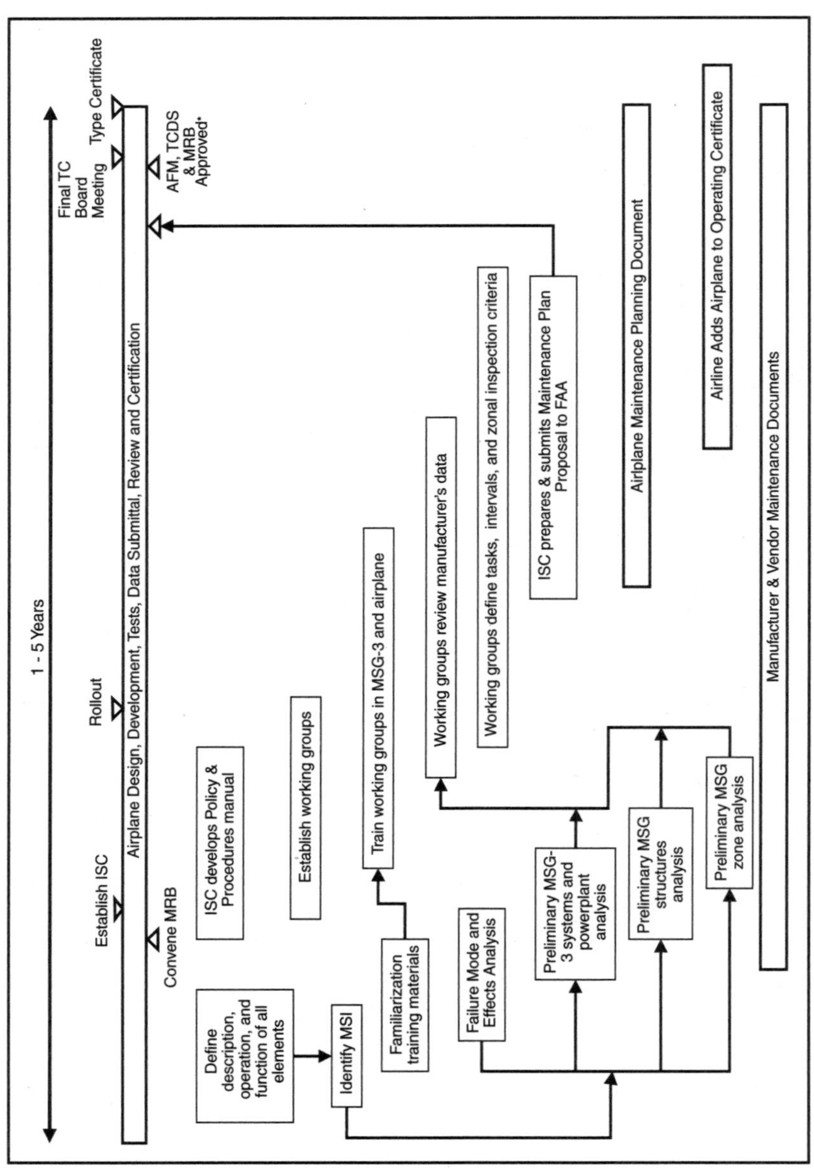

Figure 11-1 *Scheduled maintenance program development.*

- Power plant programs based on the requirements for continuing airworthiness established during power plant Type Certification
- Zonal inspection

Scheduled maintenance program development

In 1968, an ATA Maintenance Steering Group (MSG) composed of representatives of member airlines issued a handbook, MSG-1, Maintenance Evaluation and Program Development, which established a disciplined decision logic and procedures for use in developing the maintenance program for the Boeing 747.

This document was subsequently updated and made applicable for all new type aircraft. It was released as Airline/Manufacturer Maintenance Program Planning Document, MSG-2. MSG-2 decision logic was used to develop scheduled maintenance programs for the aircraft of the 1970s. MSG-2 is dominantly oriented toward maintenance processes.[1]

By 1979, with the active participation of the FAA, CAA/U.K., AEA, U.S., and European aircraft and engine manufacturers, U.S. and foreign airlines, and the U.S. Navy, the ATA revised the document to accommodate changes in technology and design and operating regulations and philosophy. It was issued as MSG-3 Maintenance Program Development.

MSG-3 currently includes three revisions, the most recent adopted in 1993. Airplanes certificated during the 1980s and 1990s use MSG-3 for the development of scheduled maintenance tasks. MSG-3 is dominantly task oriented rather than process oriented.

MSG-3 defines the physical organization of the group preparing the maintenance plan and the methodology to be followed. The group[2] tasked with developing the initial scheduled maintenance plan for a new airplane is composed of the Industry Steering Committee and the working group(s).

[1]Hard time, On-condition, and Condition monitoring. See Chapter 13.

[2]Frequently referred to as the MSG.

Industry Steering Committee

The Industry Steering Committee (ISC) establishes policy,[3] supports, and manages the activity of the working groups. Members are composed of representatives from the initial domestic and foreign purchasers of the aircraft and the airframe and engine manufacturers.

It defines the number, type, and composition of working groups and coordinates activities with the certifying FAA region, foreign regulatory agencies, and the manufacturers. It represents the interests of the airlines to the manufacturer and the regulatory agencies. A representative from the manufacturer and the airlines cochair the committee.

It documents working group proposals into a unified maintenance inspection plan. This document is supplied to the FAA. The committee gets administrative and facilities support from the manufacturer.

Working groups

Working groups identify and analyze maintenance significant items, beginning with the original list prepared by the manufacturer. The number of groups is determined by the steering committee. Usually there are separate groups for each system, that is, hydraulic, flight control, structure, power plant, and so on. Like the ISC, each working group is cochaired by a representative from the manufacturer and the airlines.

Membership in each group is composed of specialists from the interested parties. The exact size and composition of the working groups is quite flexible. It depends upon the needs of the moment as the analysis proceeds. Subgroups may be formed to resolve complex issues. The manufacturer's personnel play a large role, counseling the groups regarding details of the design.

Manufacturer's involvement

The manufacturer participates as a member of the various working groups. Additionally, before and during the formal deliberations of the working groups, manufacturers provide information concerning the component, systems, and structures design:

- Identification of an initial Maintenance Significant Items (MSI) list

[3]This is published as a policy and procedures document. This document defines the day-to-day operating details of the MSG, including the policies of the group.

- Definition of the function, operation, and unique features of the design
- Provision of failure analyses (including cause/effect), reliability data
- Preparation of a preliminary MSG-3 analysis including identification of required maintenance limitations, tasks, and intervals
- Preparation of training materials and presentation familiarization classes for the working groups and ISC

Regulatory agency representatives

FAA observers offer guidance and participate in the work of the steering committee and working groups. They are members of the FAA Maintenance Review Board (MRB). This has several advantages when the MSG formally submits its recommendation to the FAA: regulatory issues regarding the recommendations hopefully have already been resolved, time is saved discussing details of the analysis, and so forth.

Representatives from foreign regulatory agencies such as the JAA also directly participate in the process.

Airline representatives

The airline representatives to the process, as the ultimate users of the product, bring the airline maintenance expertise and perspective to the analysis. Specialists in systems, structures, and components are members of the working groups. They participate in the MSI analysis offering guidance, including recommendations for specific tasks and inspection methods.

Maintenance Significant Items (MSI)

Very early in the design, the manufacturer begins this work. A preliminary list of Line Replaceable Units (LRUs), system installations, and items of aircraft structure is drafted. The list contains items considered sufficiently significant in their maintenance requirements to justify establishing required maintenance inspections/checks. The beginning of MSI isn't very scientific. It consists of gathering a collection of old heads with gray beards and lots of experience in design and maintenance. This group looks at the preliminary design as it develops. On the basis of their collective wisdom and experience they construct this initial list. It's not rocket science.

Criteria for selection of an item as an MSI addresses the consequences of an item's failure including the effect upon safety, the undetectable failure, or the failure that has a significant operational or economic impact[4]. Each MSI must be clearly identified as to its function, functional failure, failure causes, and effects.

The list is not static. It changes as a result of design changes, reliability analyses, and failure mode and effect analyses and the working group's analyses and inputs throughout the design and certification.

MSI analysis

MSG-3 defines a methodology called *decision tree* logic. It is organized to uncover evident and hidden failures and to separate safety related to economic failure in the design. Methods for defining servicing and lubrication tasks are included. The list is analyzed for importance of the following:

- *Safety-related items.* Any system or component malfunction that results in the loss of airworthiness is by definition safety related. Is the malfunction readily apparent to mechanics or pilots? Is it hidden? The answer to these and other questions help define tasks.
- *Potential economic impacts.* The analysis examines such issues as high initial design, manufacturing and ownership cost, high maintenance cost, premature removal rates, significant access problems, and potential for mechanical dispatch delays.

Task definition

Having defined the analysis, which leads to the conclusion that a task is required, the process defines the tasks and their selection criteria.

Tasks

Tasks include:

- Lubrication/servicing, which is the replenishment of consumable materials

[4]MSG considers economic failure effect as one that does not prevent airplane operation, but is economically undesirable because of unacceptable cost of repair. Contrast this to the failure of a device included into the airplane design to assure equipment availability.

- Operational check to determine that the item is performing its intended function within quantitative tolerance
- Visual check to determine that the item is performing its intended function but does not involve quantitative tolerances
- Inspection that involves visual or specialized techniques and equipment to determine serviceability
- Functional check or quantitative check to determine serviceability of function(s)
- Restoration, the work necessary to return the item to a serviceable condition
- Discard that which involves the removal from service at a specified life limit

Task interval definition

Airplane designs assume a specific utilization and flight spectrum. The MSI analysis uses these numbers when developing the task intervals. Accumulated flight-hours,[5] calendar time, number of system operating cycles, or the number of landings are the accepted measurements used when specifying maintenance intervals.

Utilization is a significant element in determining when an inspection or maintenance task is performed. The number of flight-hours per day, the length of the flight in hours, and the number of flight cycles[6] per flight-hour all have to be considered.

Cycle-influenced items

Short flights incur more flight cycles. The number of cycles per flight-hour affects the inspections and maintenance. Cycle-influenced items include landing gear, wheels, tires and brakes, leading, and trailing edge devices.

[5]Time as measured in aircraft maintenance is either calendar based, flight time, or block time. Flight time is defined as the elapsed time between takeoff and landing. Block time is defined as the time the airplane first moves, for purposes of flight until the airplane comes to a full stop at the gate, airfreight, parking ramp, etc. Maintenance is flight time referenced; block time is crew pay and schedule related.

[6]Flight cycles refers to one completed takeoff and landing including training flight. Engine cycles refers to complete thermal cycles including the application of takeoff power. The number of engine cycles may be greater than the number of flight cycles because of touch and go training flights and/or engine run-up checks.

Flight cycles are the principal determinants of structural inspection items. They impart stresses and loads on the structure resulting in fatigue. These include gust and maneuvering loads and pressurization loads.

Engines are affected by the number of thermal cycles to which they are exposed, particularly the application of takeoff power.

Time-influenced items

Items subjected to operating wear and deterioration are obviously related to the operating hours they experience. Typical examples include systems and components installed in the airplane. Some items deteriorate not from use but merely the passage of time. These items are related to calendar time rather than airplane use; an example is emergency equipment.

Test results and reliability predictions are useful tools when selecting an interval. However, like constructing the original MSI, the process is heavily dependent upon having a good group of experienced designers and maintenance engineers to select the interval.

Maintenance recommendations

The product of the working groups' activities is a recommended list of inspection, check tasks, and check intervals. Inspection intervals may be expressed as flight time, calendar time, takeoff/landing cycles, and pressurization cycles as may be appropriate to the item analyzed.

The maintenance/inspection recommendations are assembled into a document called the Maintenance Requirements and Review Proposal. The manufacturer submits the proposal to the FAA in fulfillment of the requirements contained in 14 CFR Part 25, Appendix H. It should be noted that the proposal document is not the Maintenance Review Board report. The report will be based on the recommendations.

Maintenance Review Board

The FAA convenes the Maintenance Review Board (MRB) for examination and approval of the proposal. Final issues are identified and resolved. The board consists of representatives from the certifying Aircraft Certification Office (ACO), the Airplane Evaluation Group

(AEG), and other specialists from the Flight Standards organization. Many of the members of the board are participants in the deliberations of the working groups. The product of this review is known as the Maintenance Review Board (MRB) report.

Maintenance Review Board report

The primary purpose of an MRB report is to determine the initial scheduled maintenance requirements for new or derivative air-

DEPARTMENT OF TRANSPORTATION
FEDERAL AVIATION ADMINISTRATION
WASHINGTON, D.C.

MAINTENANCE REVIEW BOARD (MRB) REPORT
BOEING MODEL 767

REQUIREMENTS-SCHEDULED
MAINTENANCE PROGRAM

THIS DOCUMENT PRINTED BY AND AVAILABLE FROM
THE BOEING COMPANY

Figure 11-2 *Maintenance Review Board report.*

planes. It should not be confused with, or thought of as, a continuous airworthiness maintenance program. It is the framework around which each air carrier develops its own individual scheduled maintenance program for the airplane. The essential elements of the MRB are included into the airline's Operations Specifications Part D. Part D defines the maintenance program.

The airworthiness limitations identified in MRB are FAA approved and specify maintenance required under 14 CFR 43.16 and 14 CFR 91.403 or an FAA-approved program such as the air carrier maintenance program. They may not be altered, including escalation of the intervals.

Maintenance Review Board reports are valid only until the limits are published in AC 121-1A, Standard Operations Specifications—Aircraft Maintenance Handbook. This transfer of limits from the MRB report to AC 121-1A is accomplished only after sufficient satisfactory service experience has been accumulated that warrants an updating or revision of the original MRB limits.

Maintenance Planning Document

The airplane manufacturers produce the Maintenance Planning Document (MPD). It supplements the MRB. A scheduled maintenance program is constructed from the MRB report and the materials contained in the MPD. Included in the MPD are:

- Maintenance labor-hours estimates for tasks
- Facilities and tooling recommendations
- Recommended discretionary maintenance tasks improving maintenance economics, serviceability, etc.
- Administrative process and planning information, including packaging strategies

Job cards

Job cards (sometimes called task cards) translate individual inspections, checks, or other maintenance work into specific task instructions to be followed by individual mechanics when performing work. They provide space for individual sign-off by the mechanic accomplishing the work and the appropriate inspector sign-off for Required Inspection Items (RII). They are a part of the airplane maintenance record. They are divided into two categories, routine and nonroutine. Routine cards are those tasks defined by the inspection program. They have their origin in the scheduled maintenance program.

Nonroutine discrepancy cards document discrepancies discovered during the conduct of a given inspection or other maintenance activity. They are ad hoc word instructions addressing the specific repair task to be accomplished. Deficiencies discovered during the conduct of flight operations are normally not recorded on nonroutine cards. Rather the pilot in the aircraft maintenance log or the flight attendant in the cabin log enters them.

Scheduled maintenance packaging

Block maintenance

Scheduled maintenance tasks are grouped into work packages known as blocks. The exact nomenclature, composition, numbers, and sequencing of blocks varies between operators. Blocks have numerous permutations and names within the maintenance community. Understanding is further complicated by the fact that packaging concepts have evolved over the past 40+ years. Once an airplane uses a given packaging scheme, it is rarely changed to a more advanced technique. Thus the packaging program for an A-340 or 777 is different from that followed by 727/DC-8 generation airplanes. Don't let this confuse you. Regardless of its structure, the intent of any package is to level the workload, minimize airplane time out of service, and simplify the control of the tasks. Regardless of the means by which the tasks are packaged, all the required work defined by the MRB will be done when all the blocks defining the overhaul cycle are complete.

A typical block is shown in Figure 11-3. In this idealized simple illustration, each block is a multiple of the next higher block.[7] It demonstrates the concept. Each check covers all the work performed by the preceding check plus the tasks called for in the present check. Thus each succeeding check requires an increasing amount of work.

[7]In actual practice, this method of making checks exact multiples of the previous check penalizes workload scheduling.

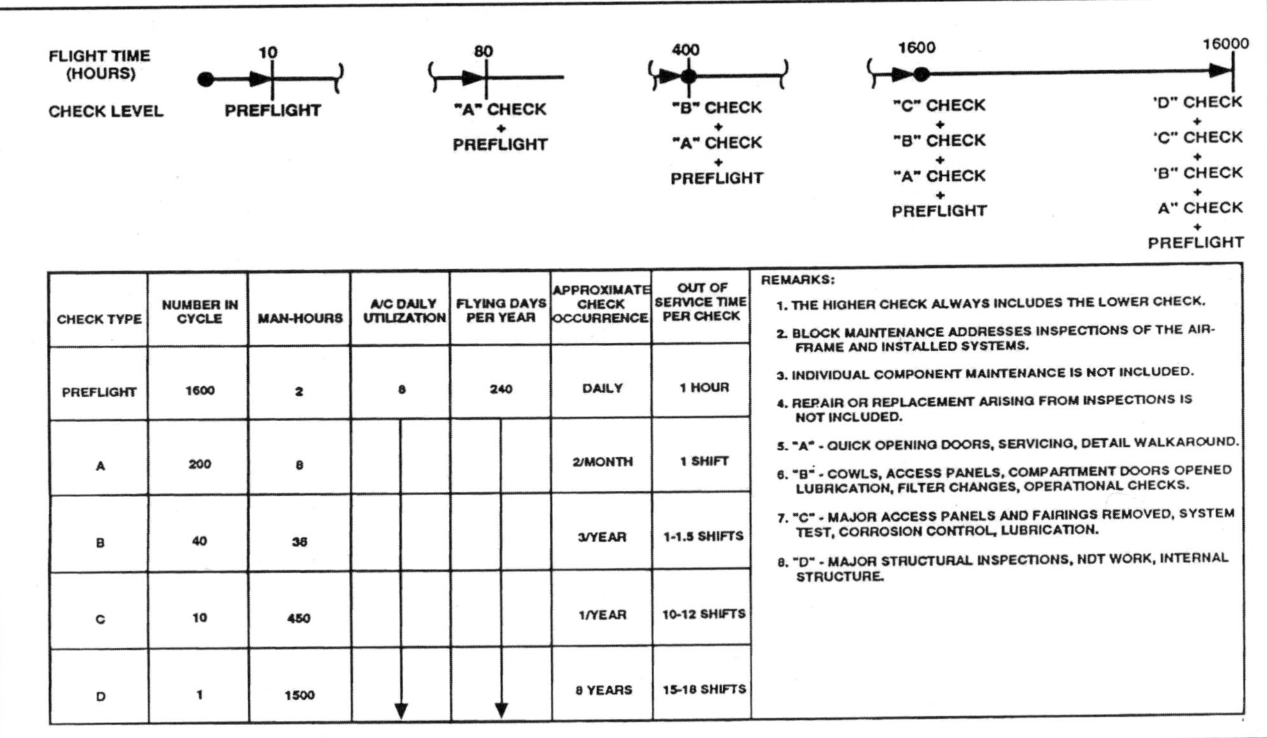

CHECK TYPE	NUMBER IN CYCLE	MAN-HOURS	A/C DAILY UTILIZATION	FLYING DAYS PER YEAR	APPROXIMATE CHECK OCCURRENCE	OUT OF SERVICE TIME PER CHECK
PREFLIGHT	1600	2	8	240	DAILY	1 HOUR
A	200	8			2/MONTH	1 SHIFT
B	40	36			3/YEAR	1-1.5 SHIFTS
C	10	450			1/YEAR	10-12 SHIFTS
D	1	1500			8 YEARS	15-18 SHIFTS

REMARKS:

1. THE HIGHER CHECK ALWAYS INCLUDES THE LOWER CHECK.
2. BLOCK MAINTENANCE ADDRESSES INSPECTIONS OF THE AIR-FRAME AND INSTALLED SYSTEMS.
3. INDIVIDUAL COMPONENT MAINTENANCE IS NOT INCLUDED.
4. REPAIR OR REPLACEMENT ARISING FROM INSPECTIONS IS NOT INCLUDED.
5. "A" - QUICK OPENING DOORS, SERVICING, DETAIL WALKAROUND.
6. "B" - COWLS, ACCESS PANELS, COMPARTMENT DOORS OPENED LUBRICATION, FILTER CHANGES, OPERATIONAL CHECKS.
7. "C" - MAJOR ACCESS PANELS AND FAIRINGS REMOVED, SYSTEM TEST, CORROSION CONTROL, LUBRICATION.
8. "D" - MAJOR STRUCTURAL INSPECTIONS, NDT WORK, INTERNAL STRUCTURE.

Figure 11-3 *Typical Block Maintenance.*

The service checks

These are the lowest levels of scheduled check. They travel under several common names: postflight, maintenance preflight, service check, overnight, number 1, number 2, to name a few. They are cursory inspections of the aircraft to look for obvious damage and deterioration. They check for "general condition and security" and review the aircraft log for discrepancies and corrective action. The accomplishment of the daily check requires little specific equipment, tools, facilities, or special skills.

It is a basic validation that the airplane remains airworthy. Usually this check will be accomplished every *X* number of days or flight-hours. There are defined specific tasks. Examples of service check items include:

- Visually inspect tailskid shock strut pop-up indicator.
- Check fluid levels.
- Check general security and cleanliness of cockpit.
- Check that emergency equipment is installed.

The letter checks

Letter checks begin to open the airplane for more detailed inspection and test. Each different letter check, A through D, is more detailed requiring more time, special tooling, special equipment, and specialists to accomplish. The C and D checks are frequently referred to as the heavy checks.

The content of each lettered check, say the C, will not necessarily be the same each time it is performed. Consider a check item that has a large interval attached to it. This item is included in a given letter check but scheduled for say only every second, third, or fourth check. For example a 2C check equals the basic C check plus those defined *X*-hour items accomplished every other C check.

The A check

This lettered check is the next highest level of scheduled maintenance. It is accomplished at a designated maintenance station in the route structure or at the main maintenance base. The check includes the daily check. It further includes the opening of access panels to check and service certain items of equipment, which are scheduled at the A check interval. Some limited special tooling, servicing, and test equipment are required. Examples of A check items include:

- General external visual inspection of aircraft structure for evidence of damage, deformation, corrosion, missing parts
- Crew oxygen system pressure check
- Emergency lights operational check
- Nose gear retract actuator lubrication
- Parking brake accumulator pressure check
- Proper operation of master warning and caution verification
- BITE test of FSEU

The B check

This is a slightly more detailed check of components/systems. Special equipment and tests may be required. It does not involve, however, detailed disassembly or removal of components. Contemporary maintenance programs do not use the B check interval. For a number of reasons, the tasks formerly defined for this interval have been distributed between the A and C checks.

The following two-letter checks are traditionally known as heavy checks. They are accomplished at the main maintenance base of the airline where specialized personnel, materials, tooling, and hangar facilities are available.

The C check

This is a detailed check of individual systems and components for serviceability and function. It requires detailed inspections and checks; a thorough visual inspection of specified areas, components, and systems; and operational or functional checks of specified components and systems. It is a high-level check that involves extensive tooling, test equipment, and special skill levels. The C check includes the lower checks, that is, the A check and the daily check. Examples of C check items include:

- Visual check of flight compartment escape ropes for condition and security
- Check of operation of DC bus tie control unit
- Visual check of the condition of entry door seals
- Operational check of flap asymmetry system
- Pressure decay check of APU fuel line shroud

- Inspection of engine inlet TAI ducting for cracks
- Operational check of RAT deployment and system

The D check

The D check, also known as the structural check, includes detailed visual and other nondestructive test inspections of the aircraft structure. It is an intense inspection of the structure for evidence of corrosion, structural deformation, cracking, and other signs of deterioration or distress. Structural checks involve extensive disassembly to gain access for inspection. Structural checks are worker-hour and calendar time intensive. Examples of structural check items include:

- Inspection of stabilizer attach bolts
- Inspection of floor beams for...
- Detailed inspection of wing box

Contemporary programs have all but eliminated the D check. These check items have been distributed among the C check packages.

Phased checks

The scheduled maintenance items for a large airplane are extensive, particularly for the higher checks, C and structural. Consequently, the accomplishment of a C check or D block removes the airplane from service for an extended period. This is unacceptable. Consider also the operator with a small fleet. It is faced with an intermittent high workload to accomplish only one or two heavy checks per year. The result is poor utilization of personnel and facilities.

A solution is to divide the C and D checks into segmented blocks or "phases."[8] This amounts to distributing, say, the C check items among the more frequent checks. Thus each lower-level check will include a different group of C check items.

Some operators, rather than distribute among the lower checks, have divided the C and D check items into a discrete number of packages that may be accomplished from night to night. Although this approach will balance the workload, it has its drawbacks. As the number of phases chosen increases, it is harder to control and record the work.

[8]This strategy is sometimes called work equalization.

- The number of times a given area of the airplane is opened up for inspection goes up unless the work division is finely tuned.
- The checks result in increased probability of maintenance error including exposing the airplane to accidental damage.
- Cumulative worker-hours expended to prepare for inspections and close the airplane go up. This is nonproductive work.
- Increased special ground-support equipment, special tooling, and test and inspection equipment must be distributed to maintenance facilities.
- More skills that are special are required to accommodate the increased number of stations performing the phases.
- Routing of airplanes becomes more complicated.

A typical phase check provides for a thorough visual inspection of specified areas, components, and systems as well as operational or functional checks of specified components and systems. Each check includes the requirements of traditional A check, multiple A check work items, and multiple portions of C and structural checks at suitable intervals. These phased intervals might vary anywhere from 200 to 800 hours, depending on the work-packaging plan and other airline operating variables.

Calendar checks

Some operators will repackage the items from a flight time–based system to a calendar time–based system on the basis of average daily usage of the equipment. The tasks are scheduled under a system of daily checks, weekly checks, and so on.

Contemporary packaging practice

Contemporary practice removes the packaging of maintenance tasks from the MRB process. When a MSG analysis is conducted only the tasks and intervals are identified. The packaging into manageable blocks is left to the operator of the airplane.

Special inspection programs

Additions to the scheduled maintenance program are augmented to accommodate specific tasks defined by such programs as Extended

Twin Operations (ETOPS).[9] Specialized navigation operations like flight operations using GPS as the sole means of navigation, or navigation checks to meet the requirements of the Category II or Category III instrument approach and landing operations require additional maintenance activity.

Aging aircraft

Supplemental structural inspection programs

High time/cycle aircraft have additional supplemental structural inspections applied to them. Supplemental structural inspection programs arose partly as the result of an accident in the mid-1970s involving a high time cycle 707 cargo airplane. Requirements for these programs are contained in a document called the Supplemental Structural Inspection Document (SSID). They are now mandated by Airworthiness Directives (AD). The SSID is a damage tolerance[10] program based on flight cycles. It focuses on fatigue cracks. Airplanes that are designed to damage tolerance requirements must have an FAA-approved Airworthiness Limitations Section as part of the Instructions for Continued Airworthiness. SSID inspection intervals defined as airworthiness limitations cannot be increased or decreased without FAA ACO approval.

During 1988 a high-cycle 737 airplane suffered major structural damage to its pressurized fuselage during flight. This directed additional activity toward detailed inspections and checks of high time structure[11]

Aging systems programs

An ARAC, known as the Airworthiness Assurance Working Group, is presently engaged to identify aging issues with aircraft systems and the possibility of requiring additional checks and inspections. Subjects being investigated include such topics as aging aircraft wire bundles, deteriorated electrical insulation, and internal leakage of hydraulic components.

[9]AC 120-42A, Extended Range Operation with Two Engine Airplanes (ETOPS).

[10]Damage tolerance is the attribute of the structure that permits it to retain its required residual strength for a period of use after the structure has sustained a given level of fatigue, corrosion, and accidental or discrete source damage.

[11]See also The Aging Aircraft Safety Act, 49 USC 44717.

Corrosion control programs

In response to the accidents just discussed and industry/FAA reviews of structural inspection practices, the FAA issued Airworthiness Directives that require corrosion prevention and control programs. These programs are calendar dependent and direct operators' inspection or maintenance programs to contain calendar-specific inspection tasks.

Low-utilization programs

Low aircraft utilization is not an air carrier issue. However, an increasing number of transport category airplanes are being used in corporate and government VIP flight operations. These airplanes have very low utilization. Consequently, the time limits and inspections set up by the Maintenance Review Board during initial Type Certification are frequently inadequate. Special low-utilization maintenance programs are available which transfer most inspections and checks from flight/cycle–based to calendar time–based intervals. Additional special inspections are also incorporated into these programs.

Check packages

The final item is to prepare a check package that bundles mandatory and discretionary maintenance tasks. Mandatory tasks include:

- The scheduled check (for example, the C check)
- AD note accomplishment
- Certification Maintenance Requirement (CMR) inspections
- Clearance of deferred maintenance MEL items
- Hard time changes of life limited parts
- Ad hoc maintenance such as corrosion control, structural repair, system repairs, component removal, and replacement
- Special operator- or manufacturer-initiated inspections

Discretionary tasks include:

- Service bulletin accomplishment to improve departure reliability

- Installation of passenger acceptance, appearance, and convenience items or cost-reduction items, for example, interior refurbishment and exterior painting
- Sampling inspections to gather data for task or check escalations, etc.
- Component replacement for convenience
- Replacement of Configuration Deviation List (CDL) items

Check package completion

The completed check package is gathered together and all the task cards, reference materials, and parts are shipped out to the hanger. The mechanics do the checks and repairs and the airplane is returned to service.

Checks are scheduled for a specific number of days. Being late out of a check disrupts the maintenance schedule. It causes a serious disruption to the maintenance plan and airplane plot. It can have a serious effect on the revenue schedule.

- Flight cancellation(s). A substitute airplane is not available or the nearest available airplane is not equipped for the particular planned flight (overwater, for example).
- Revision of the airplane schedule plot resulting in airplane substitution not of the same type. This has secondary effects:
  - Flight crew/flight attendant scheduling
  - Fleet service disruption including such items as catering, fueling, handling, and gate assignment
  - Contract commitments such as postal service contracts and DOD airlift contracts
  - Disrupted passenger seat assignment, particularly when seating assignment programs are used by the reservations group

The time allowed for a check must provide for accomplishment of all the mandatory tasks and provide time for completing unscheduled work arising from the inspection findings. Should the unscheduled work exceed the time available some work tasks must be discarded to accommodate the extra work. Consequently, a check package plan will contain a priority order for the discretionary tasks.

The priority order is not immediately intuitive until one remembers the purpose of an airline. Discretionary items are discarded in the following order:

1. *Service evaluation/inspections.* These include new or modified features that the manufacturer requests the airline to include in one or more airplanes to obtain in-service experience before committing the item to production and fleet retrofit.
2. *Cleaning/painting/refurbishment.* Exterior painting is a time-consuming project. If a given check package contains this item it is a favorite item to discard. Cleaning and refurbishment discard is a function of the airline's image. Thus, depending upon policy, it may not be allowed to discard these tasks
3. *Configuration-related items.* It is desirable to have a common airplane configuration. Among other things, it holds down spares inventory.
4. *Cost saving.* Many operators initiate projects to accomplish manufacturer service bulletins or their own modifications that are intended to reduce maintenance costs.
5. *Component sampling/convenience replacement.*
6. *Dispatch reliability improvement.* Alterations intended to improve equipment availability are only those that are having a large impact on this subject.
7. *Passenger acceptance/marketing–related items.* These are associated with interior passenger accommodations, appearance, seating configurations, passenger entertainment systems, and so on.

Further reading

Nowlan, F. Stanley, and Howard F. Heap, Reliability Centered Maintenance

FAA Advisory Circulars

AC 25.1529-1, Instructions for Continued Airworthiness of Structural Repairs on Transport Airplanes

AC 33.4-1, Instructions for Continued Airworthiness

AC 120-17A, Maintenance Control by Reliability Methods

AC 120-28D, Criteria for Approval of Category III Weather Minima for Takeoff, Landing, and Rollout

AC 120-29, Criteria for Approving Category I and Category II Landing Minima for Far 121 Operators

AC 120-33, Operational Approval of Airborne Long Range Navigation Systems for Flight within the North Atlantic Minimum Navigation Performance Specifications Airspace

AC 120-42A, Extended Range Operation with Two Engine Airplanes (ETOPS)

AC 121-1A, Standard Operations Specifications—Aircraft Maintenance Handbook

AC 121-22A, Maintenance Review Board Procedures

12

Continuous Airworthiness Programs

Introduction

Air carriers perform, or arrange with other persons to perform, maintenance, preventive maintenance, and alterations through continuous airworthiness maintenance programs.[1] This is different from other operators of aircraft. Continuous airworthiness is synonymous with maintenance. It is the activities associated with keeping the airplane in conformity to its certification.[2]

Continuous airworthiness maintenance programs (CAMP) establish the air carrier as a separate maintenance entity. Authorization to conduct a continuous maintenance program is documented in Part D of the operator's Operations Specifications.

Preventive maintenance

The term preventive maintenance has at least two meanings. The FAR define preventive maintenance as "...simple or minor preservation operations and the replacement of small standard parts not involving complex assembly operations." It is used in this context throughout the FAR.[3]

[1]14 CFR 121, Subpart L, in conjunction with 14 CFR 43 and 14 CFR 121.379, Authority to Perform and Approve Maintenance, Preventive Maintenance, and Alterations.

[2]The basis of all maintenance activity is contained in the standards of 14 CFR 91, Subpart E, and 14 CFR 43. The standards for performance of maintenance are contained in 14 CFR 43.13 and 43.15. These regulations frequently conditionally exempt air carriers operating under an approved maintenance program. Nevertheless, air carrier programs contain the elements and intent of these same regulations.

[3]14 CFR 1. Examples of preventive maintenance within this definition are contained in 14 CFR 43xA.C.

The term as used in Reliability-Centered Maintenance is "...the program of scheduled tasks necessary to insure safe and reliable operation of the equipment. The complete collection of these tasks, together with their assigned intervals, is termed the 'scheduled maintenance program.'"

Maintenance programs

Operators must substantiate a maintenance program that is acceptable to the FAA.[4] Approval and surveillance of a certificate holder's program is the responsibility of the FAA's Flight Standards Division. The assigned Principal Maintenance Inspector (PMI) administers it.

A continuous airworthiness maintenance program adequate to perform the maintenance, preventive maintenance, repairs, and alterations of aircraft is a compilation of individual maintenance and inspection functions. It contains numerous regulatory and nonregulatory elements.

The fundamental purpose of any program is the inspection, check, repair, and overhaul to ensure the continued airworthiness of:

- Airplane systems and components
- Engine and components
- Structure

Maintenance business requirements

The marketing group runs airlines not maintenance and flight operations. Marketing plans identify the revenue schedule. City pairs, airplane movement patterns, levels of service, frequency, airplane types, and number required (lift) are established by the schedule (Figure 12-1). This in turn may lead to airplane purchases, leases or equipment interchanges.

The number of airplanes available and the flight schedule dictate airplane utilization. As revenue flying increases so does scheduled maintenance frequency and consequently maintenance workload. This in turn defines facilities and administrative workloads.

[4] 14 CFR 121.367, Maintenance, Preventive Maintenance and Alterations Programs, requires that "each certificate holder shall have an inspection program..." and that the work performed is "...in accordance with the certificate holder's manual."

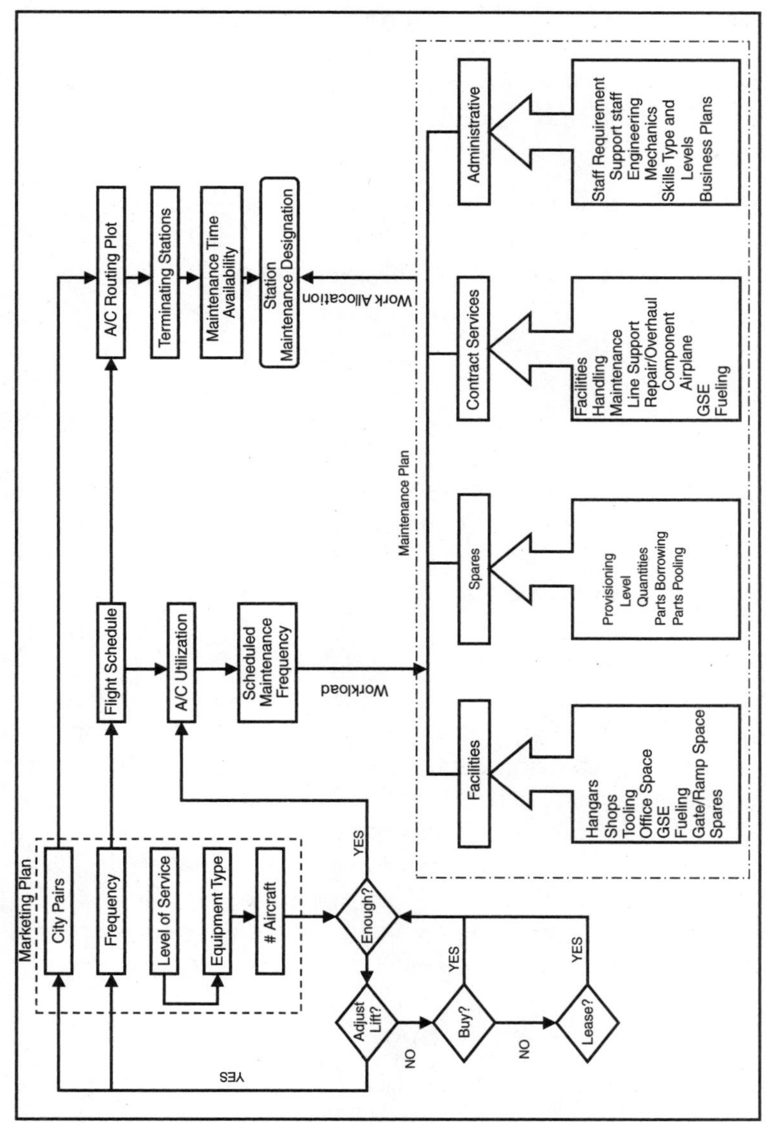

Figure 12-1 *Business requirements.*

Routing of airplanes within the flight schedule must be normalized to minimize empty nonrevenue positioning flights for schedule continuity or for maintenance. Flight departure times, daily utilization, and flight termination and layover locations are dictated by the schedule.[5] This defines the location and time available for maintenance activity. Once this is defined, then the facilities and manpower requirements will be dictated. Such things as physical plant, type of maintenance performed, staffing, and spare levels for a designated station are defined.

It should be noted that as schedule patterns and routes change, the location for maintenance activity also changes, resulting in changes to the maintenance capability of any given station.

Regulatory requirements

Numerous regulations govern maintenance operations in addition to the airworthiness requirements of the FAR. These regulations may govern the location of certain types of activity. For example, the environmental constraints imposed by some municipalities make it virtually impossible to strip and paint airplanes economically.

Maintenance program elements

Unscheduled maintenance

Unscheduled maintenance[6] is that maintenance performed to restore an item to a satisfactory condition by providing correction of a known or suspected malfunction and/or defect. Irregularities occurring during maintenance consist of failures and malfunctions, including findings from scheduled maintenance and inspections, normal flight operations, or special inspections. Unscheduled maintenance is ad hoc. It includes policies and procedures for handling discrepancies occurring during the conduct of flight operations.[7] Examples of unscheduled maintenance include:

[5]Market demands dictate this more than any factor. People don't like to fly at night or depart at the crack of dawn. They like to arrive at a time permitting them a pleasant evening and a good night's rest. They also prefer some days of the week over others as well as seasonal variations. The meaning to maintenance organizations is that mechanics work at night.

[6]Sometimes referred to as "nonroutine, nonrecurrent" maintenance.

[7]Of course, unscheduled maintenance occurs during scheduled maintenance. The inspection/checks embodying scheduled maintenance are designed to identify hidden failures that require repair or replacement.

- Resolution of aircraft pilot–generated log discrepancies (squawks)[8]
- Condition monitored failures
- Discrepancies revealed during the conduct of scheduled maintenance inspections and checks
- Special inspections initiated by the airline engineering group
- Special inspections, repairs, or replacements arising from Airworthiness Directives (ADs)
- Repair of damaged airplane structure

The nature of unscheduled maintenance dictates that it may be performed anywhere within the maintenance environment, that is, during scheduled maintenance (correction of discrepancies arising from inspections) or on the flight line (transit maintenance) while the airplane is assigned to revenue-scheduled service.

This latter environment is the most demanding. It is flight schedule driven. Time available to perform work is normally limited. Limited resources are available as well. This type of maintenance is frequently referred to as transit maintenance.

Transit maintenance. Transit maintenance[9] is that performed incidental to the normal accomplishment of the revenue schedule. It may be subdivided into through-flight or turnaround.

Through-flight maintenance, incidental to flight operations, is that performed prior to airplane departure from an en route station. Time available is usually 40 to 60 minutes. In some cases, it may be as low as 20 minutes. Equipment and personnel are limited. Work consists of a visual check of the airplane exterior with particular attention to indications of fluid leaks, obvious discrepancies such as worn or flat tires, low shock struts, fuselage or wing damage, and a review of aircraft log discrepancies. As a minimum, malfunctions affecting airworthiness are either repaired or deferred under the Minimum Equipment List (MEL) or the use of the Configuration Deviation List (CDL). The logbook is cleared.

Turnaround maintenance is performed at terminating locations in the flight schedule. Time available may be as long as 8 to 16 hours.

[8]See 14 CFR 121.563, Reporting Mechanical Irregularities, and 14 CFR 121.701(a), Maintenance Log: Aircraft.

[9]Also called turnaround.

Malfunctions affecting airworthiness are repaired. Aircraft log discrepancies are cleared. The "preflight" check will be completed. Servicing is accomplished. Such service as checking and replenishing oil supplies, lighting, cabin galleys, the water system, lavatories, the oxygen system, and fuel are completed. Passenger service items are corrected, depending on need. Interior cleaning and tidying is done.

Some scheduled maintenance (portions of a scheduled check) may be performed at stations involving a long turnaround or at a station where the aircraft remains overnight. This is also dependent upon the station's maintenance capability.

Control of deferred maintenance MEL and CDL. The control and administration of a Minimum Equipment List and Configuration Deviation List are discussed in Chapter 7.

Scheduled maintenance

Scheduled maintenance[10] programs are prepared for each airplane type operated by the airline. New type design airplanes or mature designs (new and used) may be introduced by the operator. The program includes a basic document defining the rules and procedures to be applied and a listing of tasks and intervals.

The Maintenance Review Board (MRB) report or those in the Standard Operations Specifications—Aircraft Maintenance (whichever is applicable)[11] will define recommended initial tasks and time limits. For new type designs, it will be the MRB report.

The Principal Maintenance Inspector may allow a reduction of the time limits listed in the Standard Operations Specifications—Aircraft Maintenance if an investigation of the proposed operation indicates that lower limits are advisable without compromising safety. However, higher time limits may also be imposed. The decision is guided by the following criteria:

- The operator's experience in maintaining aircraft of equal complexity
- The scope and depth of training provided for maintenance personnel on the aircraft

[10] Also called "routine recurrent" maintenance.

[11] Once a new airplane type has begun to mature in service the content and intent of the MRB is reproduced in the Standard Operations Specifications—Aircraft Maintenance.

- The intended operating environment, that is, maritime, desert, arctic, and so on
- The competency and capability of the carrier's maintenance and quality control organizations
- The adequacy of the maintenance and inspection programs
- The adequacy of all facilities and equipment for the maintenance of the airplane
- Industry service history

Individual task cards are used to provide detailed accomplishment instructions and to record accomplishment and required sign-off. Typical tasks include the replacement of life-limited items, removal of components, or overhaul for those items that have reached time limits; special nondestructive inspections; functional component and system checks/tests; and lubrication.

Overhaul and repair

This element of a maintenance program covers the activities associated with the repair and overhaul of airplanes, systems, engines, and components. It includes appropriate hangar, shop facilities, and tooling; instructions and standards for off-airplane repair and overhaul; methods for tracking, repairing, testing, approving, and recording work done on the items. Overhaul and repair may be accomplished through contract agreements with outside third-party vendors.

Maintenance manuals

The FAA requires a detailed manual system.[12] The manuals define the authorizations, methods, standards, and procedures for performance of maintenance.[13] They may take any form provided they are clearly identified, show the effective date of revision, and are easily revised. If the system consists of several independent manuals or sections, they must be available to the specific affected maintenance organizations and employees. The nonregulatory materials in the manual system should be clearly segregated from those sections required by regulations.

The manuals (Figure 12-2) define the maintenance program and provide policy, procedures, and instructions for its accomplishment. They can be divided into three broad subject categories.

[12] 14 CFR121.131 through 121.141 and 121.369.

[13] 14 CFR 43.13(c) recognizes this.

- Include instructions and information necessary to allow the personnel concerned to perform their duties and responsibilities with a high degree of safety.
- Not be contrary to any applicable federal regulation and, in the case of a flag or supplemental operation, any applicable foreign regulation, or the certificate holder's operations specifications or operating certificate.
- Content.
- General policies.
- Duties and responsibilities of each crewmember, appropriate members of the ground organization, and management personnel.
- Reference to appropriate Federal Aviation Regulations.
- Instructions and procedures for maintenance, preventive maintenance, and servicing.
- Time limitations, or standards for determining time limitations, for overhauls, inspections, and checks of airframes, engines, propellers, appliances, and emergency equipment.
- Procedures for refueling aircraft, eliminating fuel contamination, protection from fire (including electrostatic protection), and supervising and protecting passengers during refueling.
- Airworthiness inspections, including instructions covering procedures, standards, responsibilities, and authority of inspection personnel.
- Methods and procedures for maintaining the aircraft weight and center of gravity within approved limits.
- Accident notification procedures.
- Procedures and information to assist personnel to identify packages marked or labeled as containing hazardous materials and, if these materials are to be carried, stored, or handled, procedures and instructions relating to the carriage, storage, or handling of hazardous materials, including the following:
  - Procedures for determining the proper shipper certification required by 49 CFR Subchapter C, proper packaging, marking, labeling, shipping documents, compatibility of materials, and instructions on the loading, storage, and handling.
  - Notification procedures for reporting hazardous material incidents as required by 49 CFR Subchapter C.
  - Other information or instructions relating to safety.

Figure 12-2 *FAR Part 121, Required manual—maintenance content.*

- A chart or description of the certificate holder's organization.
- A list of persons with whom it has arranged for the performance of any of its required inspections, other maintenance, preventive maintenance, or alterations, including a general description of that work.
- The programs that must be followed in performing maintenance, preventive maintenance, and alterations of airplanes, including airframes, aircraft engines, propellers, appliances, emergency equipment, and parts thereof, and must include at least the following:
  - — The method of performing routine and nonroutine maintenance (other than required inspections), preventive maintenance, and alterations.
  - — A designation of the items of maintenance and alteration that must be inspected (required inspections), including at least those that could result in a failure, malfunction, or defect endangering the safe operation of the aircraft, if not performed properly or if improper parts or materials are used.
    - The method of performing required inspections and a designation by occupational title of personnel authorized to perform each required inspection.
    - Procedures for the reinspection of work performed pursuant to previous required inspection findings (buyback procedures).
    - Procedures, standards, and limits necessary for required inspections and acceptance or rejection of the items required to be inspected and for periodic inspection and calibration of precision tools, measuring devices, and test equipment.
    - Procedures to ensure that all required inspections are performed.
    - Instructions to prevent any person who performs any item of work from performing any required inspection of that work.
    - Instructions and procedures to prevent any decision of an inspector, regarding any required inspection from being countermanded by persons other than supervisory personnel of the inspection unit, or a person at that level of administration control that has overall responsibility for the management of both the required inspection functions and the other maintenance, preventive maintenance, and alterations functions.
    - Procedures to ensure that required inspections, other maintenance, preventive maintenance, and alterations that are not completed as a result of shift changes or similar work interruptions are properly completed before the aircraft is released to service.
- The certificate holder must set forth in its manual a suitable system that provides for preservation and retrieval of information in a manner acceptable to the administrator and that provides:
  - — A description of the work performed.
  - — The name of the person performing the work if the work is performed by a person outside the organization of the certificate holder.
  - — The name or other positive identification of the individual approving the work.
- Provide instruction for faster installation.
- Provide data for alignment checks.
- Structural repair manuals are FAA approved.

Figure 12-2 *FAR Part 121, Required manual—maintenance content. (Continued)*

General Maintenance Manual (GMM)

This document is developed by the operator to meet the specific requirements shown in Figure 12-2. It deals with organizational matters, the policies and procedures for the administration of the program, including planning, stores administration, and quality functions. It defines all facets of maintenance operations and their interrelationships.

The Minimum Equipment List (MEL) is usually published in the GMM (General Maintenance Manual). It is also published in various flight operations documents such as the General Operations Manual and Dispatcher Manual.

Airplane inspection program manuals

This group of publications deals with the scheduled inspection program. The information therein is derived from the FAA Maintenance Review Board Report and the recommendations of the airplane manufacturer.[14]

It includes detailed instructions (or specific references) for the accomplishment of the scheduled inspections for a particular airplane type. It normally includes the work forms or job/task cards associated with the inspections and detailed instructions for accomplishing maintenance tasks. Tasks cards may be independently developed by the operator, usually based upon manufacturer developed task cards. RII (Required Inspection Items) are usually identified on these forms.

Forms and instructions for (or references to) nonroutine tasks such as engine changes, abnormal landing inspections, and so forth, are included.

Airplane/component technical manuals

Technical manuals, principally provided by the airplane manufacturer and its vendors, address the accomplishment specific tasks. They are derived from the Instructions for Continuing Airworthiness required by FAR 25.1529 and FAR 25, Appendix H. These documents set forth methods, technical standards, measurements, operational tests, and so on. The manufacturer's material is incorporated into the operator's manual system in whole or in part by reference to it in the General Maintenance Manual. Operators frequently supplement and

[14]These are normally provided by the manufacturer in a document known as the Maintenance Planning Document (MPD).

even rewrite these manuals. It should be noted that the content of these manuals is the operator's responsibility regardless of who publishes them.

The entire manual system is enormous, characteristically consisting of several hundred thousand pages. For example, General Electric/SNECMA, a joint manufacturing engine company, annually prints 30 million pieces of paper related to engine maintenance manuals and bulletins.[15]

Electronic publishing

Increasingly, manuals are produced and distributed electronically. Current practice includes the use of CDs and the Internet and intranet applications rather than paper. Several carriers and manufacturers are implementing publishing and distribution using company intranet circuits.

Electronically produced and distributed manuals are now rapidly replacing paper systems. They assure a consistent product. They avoid the many errors that arise when paper revisions are posted into manuals. Distribution is easier. The ability to transmit this technical manual material electronically results in considerable cost saving associated with printing and distribution of materials. Additional benefits derived are the assurance of more timely and accurate data within the maintenance organization.[16]

There are, however, a number of difficulties to overcome. A large number of different, confusing systems are being offered. Each system may be convenient and cost effective for the organization selecting a publishing system, but the user is confronted with a confusing array of disparate systems requiring unique software and sometimes hardware. It is not cost effective to the user. Airlines are also selecting their own publishing systems, further compounding this problem. Integration of (or at least compatibility between) all of the available publishing systems is an elusive goal.

CD-based manuals still require manual distribution throughout the maintenance organization. Different manufacturers use different

[15]Kandebo, Stanley W., "Customer, Engine-Maker Connected by GE Web Site," *Aviation Week and Space Technology,* p. 61, April 3, 2000.

[16]It should be noted that as of this writing the FAA has not completely accepted the use of intranet-distributed documents as an acceptable form for manual materials required by 14 CFR 121.

publishing software. Internet/intranet–based systems have the advantage of overcoming distribution problems. They show great promise.

Electronic systems can very easily produce enormous quantities of data and make it available to a wide audience. This produces another problem: information overload to users. Unfortunately, many of the systems do not do a very good job of allowing users to selectively retrieve data. Rather they overwhelm users with superfluous materials in a surprising array of formats, most of which are not germane to specific user's needs. They all too often contain materials that are arbitrarily selected with a perceived notion of the end user's needs. They seem designed by publications people working with computer programmers and accountants more to solve their own problems rather than those of the end user. Don't interpret my concerns as opposition to electronic materials. Their advantages far exceed their shortcomings, which will eventually be corrected. Electronic systems are unquestionably the direction to proceed.

Return to service

A certificate holder may not operate an aircraft after maintenance unless either an airworthiness release or an aircraft log entry is executed.[17] The terms *airworthiness release, maintenance release,* and *log sign-off* are used synonymously in the air carrier world. This return to service[18] is the action identifying airplane airworthiness after the performance of maintenance.

The details of how a return to service will be accomplished are defined in the General Operations Manual. A release is accomplished by:

- A separate return to service statement—an entry referencing the return to service statement after scheduled maintenance.
- An appropriate logbook entry in the case of aircraft. This is a corrective action statement that is opposite to a discrepancy entry in the aircraft log made by the pilot in command. This entry clears the discrepancy and revalidates the original maintenance release.

[17] 14 CFR 121.709.

[18] 14 CFR 91.407, Operation after Maintenance, Preventive Maintenance, Rebuilding, or Alteration, reinforces this requirement.

- The completion of an approved serviceable tag in the case of overhaul/repair of components.

Discrepancies discovered during the performance of scheduled maintenance are usually recorded on a nonroutine job card. It provides space for a description of the deficiency as well as space to enter appropriate corrective action taken. It includes space for a sign-off by the performing mechanic (and inspector, if the item is an RII). The nonroutine job cards become a part of the work package record of the specific scheduled check being accomplished.

In case of a repair or alteration not arising from a scheduled maintenance check, a separate work package with appropriate job cards is prepared. If the repair or alteration is considered major, an FAA Form ACA 337 is included in the package.[19] The completion of the 337 constitutes a return to service. In any event, the completed work package must include a return to service statement.

Persons authorized to return to service

Persons authorized to return to service[20] after maintenance include:

- The holder of a mechanic's certificate[21] or an inspection authorization[22]
- The holder of a repair station certificate
- The holder of an air carrier operating certificate

The authorization realistically extends to only two persons, that is, certificated mechanics or mechanics with an inspection authorization. Repair stations and air carriers are agencies. Their usual agent for return to service is a mechanic or a repairperson.

Maintenance release

A release is a certification in the maintenance record that work performed on the airplane has been completed satisfactorily and

[19]14 CFR 121.707.

[20]A manufacturer may approve the return to service within certain limitations. See 14 CFR 43.7 for complete text of return to service.

[21]International Civil Aviation Organization, Annex 1, Personnel Licensing, and 14 CFR 65, Certification: Airman Other Than Flight Crewmembers.

[22]An inspection authorization is a variation of the mechanic's certificate that grants further privileges to a certificated mechanic. Persons with inspection authorizations are employed in general aviation. Their services are not required in air carrier maintenance programs.

in accordance with appropriate regulations. It constitutes one form of the return to service required by 14 CFR 43.5. The record entry reestablishes conformity, regardless of its specific format. A maintenance release certifies a complete maintenance package, for example, the C check.

Operators will list in the General Maintenance Manual when a release is required. Normally, a release is required following inspections prescribed by Part D of the Operations Specifications, maintenance activities involving Required Inspection Items (RII) inspections, or other significant maintenance.

Manner and form of the release

A release consists of a formal certification by an authorized person that:

- The work performed was in accordance with the operator's approved manual.[23]
- All Required Inspection Items (RII) were satisfactorily completed[24]
- All work was accomplished by authorized persons
- No known condition exists which renders the airplane not airworthy
- So far as the work performed is concerned, the airplane is in a safe condition.

Persons authorized to provide a release

A person specifically authorized by the operator releases the air carrier aircraft for service. The person signing a release acts as the authorized agent for the operator rather than as an individual. This form of return to service statement is a certification that the maintenance covered by the release has been accomplished according to the carrier's continuous airworthiness maintenance program. Those individuals or organizations authorized by the airline to accomplish a release are listed in the operator's General Maintenance Manual by name or organization title.

A release in no way reduces the responsibility of certificated mechanics or repairmen for maintenance they perform or supervise.

[23]The manual required by 14 CFR 121.131, 121.133, 121.135, and 121.369.

[24]14 CFR 121.3.69 (b) (2) and (b) (3) define requirements for RII within an approved maintenance program. 14 CFR 121.371 defines the requirements for required inspection personnel including limitations on such person's authority.

Responsibility for each step of the accomplished maintenance is borne by the individual signing for that step.

Validity of the release

A release is invalidated by the expiration of the required interval for a given maintenance inspection or the imposition of an Airworthiness Directive with immediate effectivity. It is correspondingly invalidated by any condition that causes the airplane to be out of conformity to its Basis of Certification.

Logbook discrepancies invalidate a maintenance release. Normally if the only required maintenance is the correction of a log discrepancy, the release is revalidated after such correction. This "log clearing" may be accomplished in one of two ways:

- The discrepancy is corrected by repair or replacement of the discrepant item with the appropriate log entry.
- The discrepancy is deferred under an approved Minimum Equipment List (MEL).

Test flight requirements

Aircraft must be test flown after maintenance or alteration that has appreciably changed the flight characteristics. However, the aircraft need not be test flown if it may be shown by suitable ground tests that the work accomplished did not appreciably affect flight characteristics.[25]

Contemporary aircraft do not require test flight after maintenance, with some minor exceptions defined by the manufacturer in the Instructions for Continuing Airworthiness. However, many operators will require test flight after maintenance as a matter of company policy. These items will usually be listed in the operators General Maintenance Manual.

Weight and balance control program

Procedures for administration of the weight and balance[26] program are contained in appropriate sections of the manual system. For example, weighing, leveling, and jacking instructions are found in the technical manuals supplied by the airplane manufacturer.

[25] 14 CFR 91.407(c).

[26] See Chapter 9.

Common weight information such as fluid weights and location, usable and unusable fluid quantities, and station locations are found in separate manuals provided by the airplane manufacturer. This subset of the maintenance program involves the periodic weighing of the airplanes as defined in a separate program for each model airplane.

Training program

Training and qualification of maintenance personnel are discussed in Chapter 13.

Safety programs

Although there is no specific 14 CFR 121 requirement for them, safety programs promote a safe environment for employees, customers, and passengers. The intent is to comply with the entire body of applicable local state and federal laws and regulations concerning health and the environment. Their significance is evidenced by the current trend to establish within the airline a separate corporate position of vice president of safety reporting directly to the chief operating officer.

Accident prevention

The program includes accident prevention policy and procedures. This includes such topics as employee safety training, safety inspections, and audits. Typical audit subjects include airfield ramp, hangar, and shop safety including outstations; fueling procedures; wildlife hazard control; maintenance safety; and hazardous materials.

Accident/disaster plans

These plans address accident/incident notification and reporting including NTSB rules.[27] Typical examples of content include:

- Reports to FAA and NTSB
- Preservation of the wreckage
- Preservation of manuals, records, and other documents relating to the airplane
- Coordination with municipal emergency services

[27] 14 CFR 121.135(20).

- Statements made to the press and public
- Participation by company employees in the accident investigation
- The disposition of cargo, mail, and security materials
- Relationships with survivors and next of kin

Hazardous materials program

These huge, complex and costly programs[28] establish and administer policy and procedures intended to:

- Meet the regulatory requirements for identification, packaging, acceptance for shipment, handling, and flight crew notification. Procedures for handling materials spills and decontamination are included.
- Assure that chemical storage, dispensing, and usage including the handling disposition of waste materials are properly managed to meet engineering, safety, health, and environmental regulations. Criteria include those established by such agencies as OSHA, EPA, aircraft and component manufacturers, and chemical vendors.

Quality assurance program

The inspection program is a quality function described in the General Maintenance Manual. It defines policies and procedures governing the conduct of inspections, the quality control organization, staffing, basic inspection policies such as inspection stamp authority, continuity of inspection, inspector authority, protection against countermand of inspector decisions, and buyback of inspection discrepancies.

Procedures for turning over inspections during work shift changes are included.[29] Unfortunately, there have been fatal accidents attributable to inadequate job turnover between shifts. Although individual work turnover is not specifically required by the regulations, prudent maintenance practice should include a system for exchanging information regarding the status of every open job remaining at shift change. This is a significant maintenance human factors issue.

[28] 49 CFR 107 and 49 CFR 171 through 175, Hazardous Materials Regulations.

[29] 14 CFR 121.369(b)(9).

Required Inspection Items (RII)

Certain items of maintenance and alteration are designated as RIIs.[30] These are defined as those maintenance and alteration items which, if not performed properly or if improper parts or materials are used, could result in a failure, malfunction, or defect endangering the safe operation of the airplane.

Typical examples of RIIs are the installation, rigging, and adjustments of flight controls and surfaces and the installation or repair of major structural components. The installation of an aircraft engine, propeller, or rotor and the overhaul or calibration of certain components such as engines, propellers, transmissions, gearboxes, or navigational equipment, the failure of which would affect the safe operation of the aircraft are also items to be considered.

The GMM identifies RIIs and contains polices and procedures for handling the inspections, including methods to prevent personnel who performed the maintenance or alteration work from also performing the required inspection of that work.

Condemned parts handling

The maintenance program should incorporate a system for the clear identification of parts that are no longer serviceable or repairable. It must assure that the materials are segregated from serviceable materials.

The FAA does not have the authority to prevent the sale or use of aircraft parts of questionable serviceability. Parts, appliances, and components from aircraft that have been involved in accidents or crashes are available to the industry as replacements. It is the operator's responsibility[31] to be aware of the possible consequences of using questionable parts on certificated aircraft. Prudent practice requires that condemned materials be physically mutilated or destroyed in order to prevent their reuse.

Continuous analysis and surveillance

This function[32] is designed to continuously audit the technical operation and personnel for conformity to regulations, technical stan-

[30] 14 CFR 121.369(b)(2) through (b)(8) and 14 CFR 121.371.

[31] Operators using a part of unknown quality, condition, or origin must be able to prove conclusively that such a part conforms to 14 CFR 43.13.

[32] 14 CFR 121.373, Continuing Analysis and Surveillance.

dards, company policies, and procedures of the maintenance program. Audits are usually conducted by the quality organization. They include surveillance:

- Of all publications and work forms to affirm that they are current and readily available to the user
- To ascertain that maintenance work is performed in accordance with the methods, standards, and techniques specified in the manual system
- To make certain that maintenance forms are screened for completeness and proper entries, and RII identification
- Of records pertaining to tracked components
- For indications of inadequate training
- To establish that airworthiness releases are properly executed by designated persons
- To determine that carryover items and deferred maintenance are properly handled
- Of the degree and frequency of adjustment and calibration of test and measuring equipment
- Of the frequency of unscheduled parts replacement or need for unscheduled maintenance
- Of safety audits
- Of audits that are designed to determine a third-party provider's level of compliance with 14 CFR Part 121, Subpart L

Maintenance reliability program

Contemporary air carrier maintenance is based on Reliability-Centered Maintenance (RCM) principles. Management of any RCM-based program requires continuing surveillance and analysis of the performance of each item under actual operating conditions. Conditions, both good and bad, arise which are representative of different types of operations. Unique maintenance practices and environments affect reliability. Reliability programs recognize this.

A reliability program is an event-reporting scheme. It involves data collection. Pilot reports, in-flight shutdowns, delays/cancellations, unscheduled engine removals, and unscheduled component removals, and so forth, are tallied, charted, and analyzed using actuarial and other techniques. The data are compared to defined performance values for airplane systems and components. The results show the effectiveness of the maintenance program and provide controls that allow adjustment of the program. Continuous monitoring and revisions assure that the appropriate type, quantity, and frequency of maintenance tasks are defined for each component, system, or appliance.

If acceptable levels of performance are exceeded no action is required. A major characteristic of reliability control programs is that they afford the operator a formal means of adjusting maintenance/inspection/overhaul intervals without prior FAA approval. This does not, however, relieve the operator or FAA of their responsibility for the effects of the program upon safety.

If performance levels are not acceptable, an investigation is initiated to assess the problem and effect corrective action. The information derived identifies failures that can effect safety and those having economic consequences. This in turn defines a need for design refinements and/or modifications to the initial maintenance program (as defined in the MRB report) including task interval adjustments.

Airplane components and systems that do not experience a statistically significant number of repeatable events are monitored as well. Surveillance of premature removal rates of components, operational difficulties attributable to maintenance such as delay causes, excessive MEL deferral of components or systems, unusual parts or materials usage, number of pilot reports is conducted. Subjective performance levels for these items are established, many times derived from empirical evidence. The system provides for timely identification and correction of discrepancies or emerging problems before they become significant.

Reliability programs provide a useful design evaluation for future designs. This includes:

- Information defining the failure modes of components removed as a result of failure
- The condition of serviceable parts in components that have failed
- The condition of serviceable units inspected as samples

Programs are rooted in an information system consisting of:

- An effective data collection system.

- A system for effectively assessing the need for changes in maintenance interval or design and for taking appropriate action. Action consists of appropriate reviews of the following:
  - Actuarial or engineering studies employed to determine need for maintenance program changes.
  - Actual maintenance program changes involving inspection frequency and content, functional checks, overhaul limits, and times.
  - Aircraft, aircraft system or component modification, or repair.
  - Other actions peculiar to the operating conditions that prevail.
  - Preparation of appropriate reports.

The operator uses appropriate items from the following performance data as the basic elements of the program:

- Unscheduled removals
- Confirmed failures
- Deficiencies observed and corrected but not otherwise reportable
- Pilot reports
- Sampling inspection
- Functional checks
- Shop findings
- Bench checks
- Mechanical reliability reports
- Mechanical interruption summary reports

Spares program

There must be adequate spares available. The program must include such topics as receiving, handling, and guaranteeing that all materials purchased for maintenance are properly documented to assure their individual airworthiness. Proper storage and handling facilities must be available including, as appropriate, materials quarantine areas to assure segregation of serviceable, repairable, and condemned items.

Suitable materials identification procedures must be established, including a tagging system.

Part borrowing and pooling

Part borrowing requires specific FAA approval by the addition of an authorization to the Operations Specifications. The specifications will require the operator to have procedures that:

- Indicate the parts are clearly established as conforming to the Basis of Certification or manufactured by an air carrier or a repair station under its approved program. This includes parts manufactured under a PMA or provided by the Type Certificate holder. Parts must display evidence of FAA approval, verifying origin and serviceability.
- Restrict the overhaul time limits to those authorized by Operations Specifications.
- Restrict the remaining minimum time to overhaul to that authorized by Operations Specifications.
- Ensure the operator has an approved list of authorized vendors, repair stations, and air carriers from which it may borrow parts.

Parts borrowing includes a short-term authorization to use a part that does not conform to the operator's program, that is, having a higher time-in-service limit than the air carrier's currently approved limit.

Part pooling is a form or borrowing. High-cost components like engines are frequently pooled to control costs. One member of the pool will maintain and store the component at specified locations. Responsibility for airworthiness of the pool component normally rests with the holder of the pool component. The participants in the pool must ensure that the interchangeability of the pool part is assured for all members of the pool.

Unapproved parts

Suspected Unapproved Parts (SUP) programs address new parts of indeterminate origin and certification as well as salvaged parts.

Unapproved parts are those that do not have a clear certification trail from a manufacturer or holder of a PMA. Many of these parts are counterfeit. They frequently are made from the unapproved materials, standards, and manufacturing techniques. In the last 10 years, this activity has increased alarmingly.

Salvaged parts have frequently been subjected to forces or environments that exceed the design specifications. There is evidence that an increasing numbers of parts salvaged from crashed airplanes, or in some cases stolen, are entering the spares market.

Cold weather operation program

Special attention is directed toward cold weather operations including snow removal,[33] ground deicing, and anti-icing.[34]

Fuel control program

Fuel programs, providing procedures for control and quality of engine fuel, are normally administered by maintenance organizations. Refueling procedures[35] frequently are contained in a separate document made available to the company and contract fueling personnel.

Reporting requirements

Numerous reports arise in the normal course of business. However, the following reports are specifically required by the FAR.

Mechanical reliability reports

Like manufacturers, air carriers are required to report[36] certain malfunctions, failures, or defects occurring to their aircraft every 24 hours to the Principal Maintenance Inspector. See Chapter 5. The reports are combined with reports from other operators and from organizations required to report the same type of items required by 14 CFR 21.3.

[33]Required of certificated airports. See 14 CFR 139.313, Snow and Ice Control. "Each certificate holder whose airport is located where snow and icing conditions regularly occur shall prepare, maintain, and carry out a snow and ice control plan."

[34]14 CFR 121.135(b)(14).

[35]14 CFR 121.135(b)(18).

[36]14 CFR 121.703, Mechanical Reliability Reports.

Mechanical interruption summary report

In addition to the reports required above, operators must report[37] the following occurrences to the Administrator.

- Each interruption to a flight, unscheduled change of aircraft en route, or unscheduled stop or diversion from a route, caused by known or suspected mechanical difficulties or malfunctions that are not required to be reported under 14 CFR 121.703.
- The number of engines removed prematurely because of malfunction, failure, or defect, listed by make and model and the aircraft type in which it was installed.
- The number of propeller featherings in flight, listed by type of propeller and engine and aircraft on which it was installed. Propeller featherings for training, demonstration, or flight check purposes need not be reported.

Alteration and repair reports

A report[38] of each major alteration or major repair of an airframe, aircraft engine, propeller, or appliance of the aircraft operated must be completed and kept.

Records system

14 CFR 43.9(b) requires "...(the) holder of an air carrier operating certificate...required by its...operations specifications to provide...a continuous airworthiness maintenance program, shall make a record of the maintenance, preventive maintenance, rebuilding, and alteration, on aircraft, airframes, aircraft engines, propellers, appliances, or component parts which it operates."

A record system ensures that adequate and appropriate records are maintained for all work performed on airplanes and are readily available. Any records system should classify all records into retention categories to improve storage and retrieval of data. Representative categories are:

- *Continuous.* These records are continuously updated. These include such records as aircraft, power plant, or component time in service.

[37]14 CFR 121.705, Mechanical Interruption Summary Report.

[38]14 CFR 121.707, Alteration and Repair Reports.

- *Routine.* Time-limited records, which are maintained for a period of *X* months. Examples include records of maintenance releases not associated with overhaul.
- *Repetitive.* Work that is repeated at specific intervals. This includes scheduled maintenance check records for at least one complete overhaul cycle.
- *Permanent.* This category includes permanent changes to the configuration of an airplane, engine, component, and/or appliances classified as major repairs and alterations, and the airworthiness directive compliance record. Such records are retained permanently. It also includes records of any FAA approvals, actions, directions, or exemptions.

Any record system adopted must clearly segregate records required by law and regulation from those associated with the normal conduct of business. Records required by law or regulation must be made available to authorized representatives of the FAA or the NTSB.[39]

For each record maintained, required, or business related, a retention period must be defined and studiously followed. Do not allow records to be retained past their required retention date. Record storage and processing is expensive. Even invalid or out-of-date records can be subpoenaed during legal discovery proceedings associated with litigation. A record all FAA approvals, directions, and exemptions must be fully documented and retained. It is best to not retain drafts, notes, and diaries as a part of any record. This includes personal records.

Airplane records

Airplane records[40] include the following elements:

- Records necessary to show that all requirements for the issuance of an airworthiness release[41] have been met. These are retained until the work is repeated or superseded by other work or for 1 year[42] after the work is performed.

[39]14 CFR 121.380(d), Maintenance Recording Requirements.

[40]14 CFR 121.380, Maintenance Recording Requirements.

[41]Required by 14 CFR 121.709, Airworthiness Release or Aircraft Log Entry.

[42]Except the records of the last complete overhaul of each airframe, engine propeller, or appliance.

- The records of the last complete overhaul of each airframe, engine, propeller, and appliance must be retained until the work is superseded by work of equivalent scope and detail.
- Records indicating total time in service of the airframe, engine, and propeller.
- Records showing the current inspection status of the airplane.
- Record of the current status of life-limited parts including the time since last overhaul.
- Record of major repairs and alterations contained in FAA Form ACA 337.
- Record of Airworthiness Directive compliance.
- Configuration control records.

When the airplane is sold these records transfer[43] with it to the new owner.

Component records

Component records are kept by the individual repair shops accomplishing work on rotatable units. They include such data as service bulletin, AD note status, software load versions, time/cycles in service, and total time/cycles. Components should be tracked as they enter the airline inventory and move from stores to airplane to shop throughout their lives. Most tracking is accomplished by a component tracking system.

Component tag system

Rotatable units, whether held in stores, shops, flight kits, or line station holding areas, must be identified as to their serviceability. The most common method is a component tagging system. There are numerous schemes used. Whatever the system, tags used to identify components fall within several classifications:

- *Serviceable tags* are used to identify both serviceable and unserviceable rotatable units.
- *Unserviceable tags* identify units which are not serviceable but are repairable.
- *Repairable tags* are used to identify rejected and rotatable units.

[43]14 CFR 121.380a, Transfer of Maintenance Records.

- *Condemned tags* identify units that are determined to be unserviceable and unrepairable.

Condemned property

Property no longer serviceable or repairable must be clearly segregated from repairable or serviceable materials. Procedures should be in place to assure that no condemned materials will be reintroduced into the stores/maintenance chain.

Additionally procedures should include destruction or disfigurement beyond restoration of all condemned parts. This assures that the materials are not reintroduced into the system by unscrupulous salvage operations.

Parts robbing

Parts robbing is, in general, poor practice and should be avoided. However, it is sometimes necessary. Parts robbing must be carefully tracked and recorded. Parts rob tags are used to transfer a serviceable unit from one airplane or higher assembly to another.

Return to service

Operators must retain all of the records necessary to show that all requirements for the issuance of an airworthiness release have been met.

Logbooks

Discrepancies occurring during line operations are recorded in the operator's logbook system. Logbook procedures are contained in the General Maintenance Manual. Logbook systems usually consist of three basic bound volumes carried on board the airplane:

Maintenance log. This log is used to record discrepancies encountered during operation of the airplane and the actions taken to correct the discrepancies. This log may also include space to record additional flight data including names of crewmembers, flight parameters, payload data, fueling data, and flight hours and cycles operated.[44]

14 CFR 121.701 requires that a certificate holder have an airplane maintenance log. This log is one element of the required maintenance records. It contains the record of actions taken to correct observed failures or malfunction reported by the pilot in command.[45]

[44]Some operators record this information in a separate log commonly called the "flight log" or "journey log."

[45]14 CFR 121.563, Reporting Mechanical Irregularities.

The log must be accessible to the flight crew. The pilot in command must review the log before flight to determine the status of previously recorded discrepancies.

Cabin discrepancy log. Cabin logs are used by cabin attendants to record nonairworthiness discrepancies encountered in the passenger cabin and corrective actions taken by maintenance. They are used to avoid filling the required maintenance log with items not required to be recorded by regulation. When using a cabin log it is important that the procedures covering its use include review by qualified personnel to make sure that airworthiness items are not entered.

Deferred maintenance item (DMI) log. MEL-deferred items will be listed in the maintenance log on numerous individual pages. This makes it difficult for the pilots and mechanics to determine the status of the airplane. The DMI log provides flight and maintenance personnel with a convenient single source for review of deferred items. See Chapter 4 for further details.

Airplane listing

Certificate holders conducting domestic or flag operations[46] must maintain a current list of each aircraft that they operate in scheduled air transportation.

Airplane configuration control

Any maintenance program must include a means for tracking the configuration of each airplane. This may take any form. At the time of delivery of the airplane the manufacturer provides an equipment listing that identifies by name and specific part number (including modification status) each item of installed equipment. This list should be continued throughout the life of the airplane. Configuration control must include the incorporation status of the airplane and installed components, and also:

- Any repairs or alterations, not just major repairs or alterations
- Manufacturer service bulletins and other service documents
- Installed computer software
- The AD note compliance record[47]

[46] 14 CFR 121.685, Aircraft Record: Domestic and Flag Operations.

[47] 14 CFR 121.380(2)(vi).

Maintenance operations

Maintenance operations may be divided into line maintenance and base maintenance disciplines and support functions.

Line maintenance

This is that maintenance activity performed while the airplane is committed to the revenue schedule. It is the principal determinant of mechanical schedule reliability. It is essentially nonroutine, nonrecurrent maintenance. It may be subdivided into gate or turnaround.

Gate maintenance is performed before airplane departure, either originating a flight or handling a transit flight. It is incidental to scheduled flight operations. Time available is normally limited, usually 40 to 60 minutes, but may be as low as 10 to 20 minutes. Equipment and personnel are also limited. Work characteristically consists of a visual check of the airplane exterior with particular attention to indications of fuel or oil leaks, obvious discrepancies such as worn or flat tires, low shock struts, and fuselage or wing damage.

Only that work which is necessary to keep the airplane airworthy without incurring a flight delay or cancellation is accomplished. Gate maintenance is not there to make the airplane pretty, just safe and moving.

The aircraft log is reviewed and cleared. At a minimum, malfunctions affecting airworthiness are either repaired or deferred under the Minimum Equipment List (MEL). Only simple uncomplicated in situ repairs are accomplished. Removal and replacement of line replaceable units (LRU) are only accomplished if required.[48]

Turnaround maintenance, also known as overnight maintenance, is performed at terminating locations in the flight schedule. Time available may be 8 to 16 hours or more. The environment is less de-

[48]Gate mechanics must be able to "think on their feet." Proficiency in a number of maintenance disciplines including knowledge of the entire airplane and its systems is required. They live in the most intense, schedule-driven, results-oriented environment in aviation. Theirs is an ad hoc world. It is impossible to plan failures on inbound airplanes so that mechanics can be prepared. There is little time for deliberation or research of problems. They have the privilege of working in the rain or snow on a windy ramp at all hours of the day and night including holidays and Sundays. Gate mechanics must have a certain "bedside manner" because they are dealing directly with people outside the maintenance community, who frequently do not have the expertise to describe the problem or who frequently have a separate agenda.

manding than when the airplane is committed to the revenue schedule. The work is predictable. Many times it is accomplished by line maintenance mechanics.

The work usually consists of a daily check. Logbook discrepancies, including MEL deferrals, are corrected. Passenger service equipment discrepancies are corrected. Servicing is accomplished.

Depending upon time and personnel availability additional low-level scheduled maintenance (for example, portions of a phased check) may be performed at stations having a long turnaround. Discretionary tasks may be included. AD note accomplishment may also be included in the work package.

Base maintenance

Base maintenance is divided into three distinct categories—hangar, shop, and support activity.

Hangar maintenance is affiliated with work on the airplane when it is removed from the revenue schedule. It is predominately, though not exclusively, associated with heavy scheduled maintenance checks (C and D); AD accomplishment, the incorporation of aircraft alterations, structural repairs, corrosion control tasks, and discretionary work. Hangar maintenance is also predictable. It is performed in a benign atmosphere. Many times it is in a nice warm hangar rather than outside exposed to the adverse weather common to many stations. There is a defined work package. Maintenance arising from discrepancies that are found during inspection is less demanding because time is allocated in the work package to correct discrepancies. Heavy checks may take from 5 to 25 days.

Shop maintenance consists of repair, overhaul, or refurbishment of components or major assemblies (for example, power plants) which have been removed from the airplane. Usually work is confined to one or more LRUs or at least to a given specialty (for example, sheet metal work, accessory overhaul). There is a predictable workload.

Support functions

Support functions consist of normal business activities such as planning, scheduling, work control, maintenance control, engineering, and training as well as conventional administrative activity.

Airplane ownership

Who owns the airplanes? Rules of ownership vary with each operator. It is characteristically divided among maintenance, operations control, flight operations, passenger services, and fleet services. Ownership delineates control responsibility within the operator's organization. It affects delay reporting.

Maintenance ownership

Maintenance owns the airplane whenever it is released from the revenue schedule by operations control for scheduled maintenance at the completion of a given flight or flight segment until required maintenance has been completed and the airplane has a valid maintenance release. Maintenance also owns the airplane whenever the airplane is rejected by the pilot in command for maintenance reasons, until the problem is corrected. Maintenance returns ownership to other organizations upon completion of any required maintenance task and the execution or revalidation of a maintenance release to return the airplane to service. Ownership may be further divided between two principle groups within the organization, line maintenance and base maintenance

Maintenance facilities

Certificate holders must show as a part of their certification that they have adequate facilities including communications capability, people, physical plant, tooling, and spares to support the maintenance of the airplanes. This includes facilities at locations along the route structure. These may be leased or contracted facilities and services. As has been seen, the locations must be carefully chosen.

Line maintenance station classification

Not every station within the operator's route structure will have maintenance capability. In fact, the majority do not. Many stations have a limited number of flights arriving. The location and level of maintenance capability is a business decision. Adequate maintenance facilities, however, remain a regulatory requirement and thus are subject to FAA monitoring.

Should an airplane encounter difficulties at a nonmaintenance station, company personnel may be flown to the airplane. Alternative solutions include either preexecuted or ad hoc support contracts with other operators at the station in question. The least desirable solution is to obtain a special flight permit to ferry the airplane to a maintenance station.

As schedule, patterns, airplane assignments, and routing change, station designations will change. Some will become new maintenance stations; some will close. Capability and resources will change.

Once designated as a maintenance station the level of work the station will perform is defined. This in turn defines the staffing level, tooling, spares levels, and the like. A typical station classification list is shown in Figure 12-3.

Third-party maintenance

Operators frequently contract the services of third parties to accomplish all or part of their continuous airworthiness maintenance programs. Contractual arrangements extend the maintenance facilities and capabilities of an operator. In effect, they are an extension of the operator's maintenance program.[49] Type and amount varies from airline to airline.

- Business pressures demand efficiency and quality wherever they can get it.
- Scale is the first criterion.
- Efficiency and the ability to specialize are the next considerations.
- High-tech and special repairs are farmed out.
- Balancing workload can also mean doing work for others.
- One-of-a-kind work packages are farmed out to avoid scheduling, regulatory requirements work leveling, labor problems.
- Older airplanes are being farmed out.
- Outsourcing is increasingly restrained by union considerations. Some still do all the work on their fleets as long as capacity is available.
- Labor is getting scarce.

[49]The 10 largest U.S. passenger airlines outsourced $2.43 billion in maintenance repair and overhaul work in the 12 months ended September 30, 1999, a 10-fold increase since 1985. This represents 21.8 percent of their total maintenance budgets.

Station	Category*	Model	Preflight	Overnight	A Check	B Check	C Check	D Check
SFO	1	All	✓	✓	✓	✓	✓	✓
ORD	2	727	✓	✓	✓	✓	✓	
		737	✓	✓	✓	✓	✓	
		777	✓	✓	✓	✓	✓	
		A340	✓	✓	✓	✓	✓	
SEA	3	L-1011	✓	✓	✓	✓		
AKL	4	777	✓	✓				
BKK	5	747	✓					
YVG	6	727						

*Category 1: Full repair and overhaul capability of airframe, power plant, and components, including machine shop repair, plating, sheet metal and composite repair, avionics/electrical components, mechanical and electrical instrument repair, NDT capability, and complete spares support.

Category 2: Full repair and overhaul of airframe (usually a specific model), limited component repair, spares, NDT, etc.

Category 3: Ability to accomplish minor repairs to airframes and power plants. No component repair capability. Limited spares and shops.

Category 4: Limited to simple remove and replace of components and system servicing. No shop capability. Limited spares.

Category 5: No maintenance capability or spares. Minor system servicing such as fueling, oil service.

Category 6: Limited maintenance, servicing, and spares provided by designated contract agent. Ability to remove and replace components.

Figure 12-3 *Maintenance station classification.*

The operator must list in its manual[50] the persons with whom it contracts for maintenance and include a description of the contracted work. The contract and details of the obligations/commitments of each participant must be referenced on Operations Specifications. The Operations Specification provides for cancellation. Increasing pressures from regulators demanding more control and documentation can favor bringing some work in-house or consolidating third-party work with big vendors.

Operators contracting to have maintenance performed by another certificate holder may be authorized by their Operations Specifications to adopt all or part of the contractor's maintenance program, including involvement in that contractor's reliability program.

Maintenance contracts are becoming more and more the norm. They are also becoming increasingly complex and convoluted. Individual contractors may subcontract work to other maintenance agencies. It is thus frequently difficult to assure that the work accomplished complies with the operator's maintenance program. Nevertheless, the operator retains primary airworthiness responsibility regardless of the terms of any contractual arrangement. Maintenance agreements must clearly define the specifications and standards required for accomplishing work. It is incumbent upon the operator to determine the capability of third parties, including the subcontractors who do the work and the suitability of the arrangements. This includes oversight to ensure that the work contracted is accomplished in accordance with the operator's approved program.

Detailed technical and administrative content of contracts is contained in the operator's General Maintenance Manual. The maintenance manual of the operator must describe the policies and procedures for administering the contractual arrangement. The appropriate manual and technical material must be provided to the contractor. However, the operator may adopt the publications of a contractor in part or in total.

Contract maintenance categorization

The most common contract agents are repair stations or other air carriers. Work contracted may consist of simple ground handling, fueling, and line maintenance at various stations along the operators

[50]14 CFR 121.369(c)(2).

route structure or detailed scheduled inspections and overhaul/repair work with large maintenance providers. In the extreme, all maintenance may be contracted. The FAA categorizes contract maintenance service as follows.

Category A

Within this classification, the operator arranges for the performance of specific maintenance work. It includes contracts with repair stations, certificated mechanics, or other certificated operators to repair, inspect, or overhaul engines, structures, airframes, and/or appliances. Work is accomplished in accordance with the operator's approved program. The operator's manual must list the names of these organizations and the scope of the work contracted. But specific Operations Specifications approval is not required.

It is common for events to occur away from an operator's normal maintenance locations. Typical examples include unscheduled engine changes, structural repairs, maintenance on airplanes diverted to a station for technical or weather reasons, and so on.

Category B

In this type of arrangement, the operator contracts a total maintenance program. All maintenance is performed in accordance with the contractor's programs, methods, procedures, and standards. The operator's equipment is considered part of the contractor's fleet for purposes of maintenance program content and maintenance intervals, including reliability control. Reliability data generated by the operator's equipment must be accounted for in the contractor's reliability program, unless data generated by the contractor's fleet are adequate and appropriate to the operator's fleet. Data generated by the operator's fleet should be compared periodically to data from the contractor's fleet. The contractor must account for all inconsistencies. This type of contractual arrangement must be authorized by Operations Specifications

Maintenance program content changes and interval adjustments may not require approval by the Principal Airworthiness Inspector. If approval is required, the operator is obligated to provide the Principal Airworthiness Inspector(s) with supporting data on which such changes are based.

The operator generally is approved for the contractor's existing maintenance intervals. Special requirements may be needed to compensate for configuration differences, operational and environmental conditions (geographic areas, etc.), or other variables (hours per cycle versus cycles per hour). The maintenance time-limitation section of the Operations Specifications must identify any special requirements, either specifically or by reference to another document approved by the Administrator.

Category C

In this classification, the operator contracts specific functions using the contractor's approved maintenance program. This category is similar to Category B except that the contract covers specific functions rather than an all-encompassing program. For example, the contract may cover heavy maintenance on engines under the contractor's approved maintenance program. These contracts shall be approved for use by Operations Specifications.

Category D

In this arrangement, the operator contracts to participate in the contractor's FAA-approved reliability program. In this category, the operator does not use the contractor's maintenance program, but participates in the contractor's FAA-approved reliability program. This type of contractual arrangement may encompass the entire aircraft, or engines, and must be included in the contractor's fleet for reliability purposes. This arrangement must be approved for use by Operations Specifications.

Substantial maintenance providers

Contractors, who perform substantial maintenance, must be authorized and listed on Operations Specifications. Their compliance with 14 CFR Part 121, Subpart L, must be confirmed. All new maintenance providers must be authorized and listed on Operations Specifications.

Repair stations

Repair stations are agencies that perform maintenance, repair, and alteration of aeronautical products. They are broadly classified as domestic and foreign. Like manufacturers and air carriers, the FAA must also certificate repair stations.

Certification consists of showing that the applicant for a certificate has sufficient personnel, facilities, materials, and management personnel to accomplish the work for which they are certificated. Appropriate Operations Specifications are issued to the approved agency. A domestic repair station certificate is indefinite, whereas a foreign certificate is limited to 12 months

The rating and class of work that may be performed by a certificated domestic repair station are shown in Figure 12-4. Limited ratings authorizing work on a specific make/model of equipment may also be granted.

The extent of the work done by a repair station runs from organizations certificated for a single specific item to so-called full-service organizations that run the gamut from simple mom and pop operations to full-service facilities. Repair stations do airframe and component inspection repair and overhaul. Airlines and manufacturers may also have a repair station certificate.

Amendment of the maintenance program

General

Routine changes to the program such as interval adjustments or the content of the manual system may not require specific FAA approval of the changes. However, the Principal Maintenance Inspector will monitor this activity.

The amendment of a certificate holder's maintenance program requiring extensive revision and in some cases specific FAA approval is caused by many factors. Notable examples include:

- The addition of an airplane into the certificate holder's fleet
- Increases in time intervals for scheduled inspection items arising from reliability programs (escalation)
- Aircraft leases, interchange agreements
- Changes to third-party service and maintenance providers
- Parts borrowing and pooling arrangements

Adding airplanes to program

The addition of a new airplane type into a maintenance program is straightforward, particularly a new type design which has not proved itself in service. Under these circumstances, the amendment consists of incorporating the 777 maintenance MRB report into the maintenance program and the Operations Specifications.

Domestic Repair Station Rating		
Rating	**Class**	**Authorization**
Airframe	1	Composite construction of small aircraft
	2	Composite construction of large aircraft
	3	All metal construction of small aircraft
	4	All metal construction of large aircraft
Power plant	1	Reciprocating engines of 400 horsepower or less
	2	Reciprocating engines of more than 400 horsepower
	3	Turbine engines
Propeller	1	All fixed pitch and ground adjustable propellers of wood, metal, or composite construction
	2	All other propellers, by make
Radio	1	Communication equipment: Any radio transmitting equipment or receiving equipment, or both, used in aircraft to send or receive communications in flight, regardless of carrier frequency or type of modulation used, including auxiliary and related aircraft interphone systems, amplifier systems, electrical or electronic intercrew signaling devices, and similar equipment, but not including equipment used for navigation of the aircraft or as an aid to navigation, equipment for measuring altitude or terrain clearance, other measuring equipment operated on radio or radar principles, or mechanical, electrical, gyroscopic, or electronic instruments that are a part of communications radio equipment.
	2	Navigational equipment: Any radio system used in aircraft for en route or approach navigation, except equipment operated on radar or pulsed radio frequency principles, but not including equipment for measuring altitude or terrain clearance or other distance equipment operated on radar or pulsed radio frequency principles.
	3	Radar equipment: Any aircraft electronic system operated on radar or pulsed radio frequency principles.

Figure 12-4 *Repair station class ratings.*

Domestic Repair Station Rating		
Rating	**Class**	**Authorization**
Instrument	1	Mechanical: Any diaphragm, bourdon tube, aneroid, optical, or mechanically driven centrifugal instrument that is used on aircraft or to operate aircraft, including tachometers, airspeed indicators, pressure gauges, drift sights, magnetic compasses, altimeters, or similar mechanical instruments
	2	Electrical: Any self-synchronous and electrical indicating instruments and systems, including remote indicating instruments, cylinder head temperature gauges, or similar electrical instruments
	3	Any instrument or system using gyroscopic principles and motivated by air pressure or electrical energy, including automatic pilot control units, turn and bank indicators, directional gyros, and their parts, and flux gate and gyrosyn compasses
	4	Electronic: Any instruments whose operation depends on electron tubes, transistors, or similar devices including capacitance-type quantity gauges, system amplifiers, and engine analyzers
Accessory	1	Mechanical accessories that depend on friction, hydraulics, mechanical linkage, or pneumatic pressure for operation, including aircraft wheel brakes, mechanically driven pumps, carburetors, aircraft wheel assemblies, shock absorber struts, and hydraulic servo units
	2	Electrical accessories that depend on electrical energy for their operation, and generators, including starters, voltage regulators, electric motors, electrically driven fuel pumps, magnetos, or similar electrical accessories
	3	Electronic accessories that depend on the use of an electron tube transistor, or similar device, including supercharger, temperature, air-conditioning controls, or similar electronic controls

Figure 12-4 *Repair station class ratings. (Continued)*

Domestic Repair Station Rating
Limited Ratings
Limited ratings are issued for (1) airframes of a particular make and model; (2) engines of a particular make and model; (3) propellers of a particular make and model; (4) instruments of a particular make and model; (5) radio equipment of a particular make and model; (6) accessories of a particular make and model; (7) landing gear components; (8) floats, by make; (9) nondestructive inspection, testing, and processing; (10) emergency equipment; (11) rotor blades, by make and model; (12) aircraft fabric work; and (13) any other purpose for which the administrator finds the applicant's request is appropriate. Manufacturers may hold a limited repair station certificate for: (1) Aircraft manufactured by the holder of the rating under a Type Certificate or a Production Certificate (2) Aircraft engines manufactured by the holder of the rating under a Type Certificate or a Production Certificate (3) Propellers manufactured by the holder of the rating under a Type Certificate or a Production Certificate (4) Appliances manufactured by the holder of the rating (i) under a Type Certificate, (ii) under a Production Certificate, (iii) under a TSO authorization, or (iv) in accordance with § 21.303 of this chapter. (5) Parts manufactured by the holder of the rating under a TSO authorization or in accordance

Figure 12-4 *Repair station class ratings. (Continued)*

Appropriate changes to facilities, quality programs, and so forth, are accomplished.

However, the MRB report is a generic listing of maintenance tasks without regard either to airplane configuration or to the areas, routes, and types of operations in which a specific carrier will operate the airplane. Therefore, a maintenance program may be initially adjusted over the MRB requirements to meet these unique air carrier environments. They arise from the reliability program. The result may be more inspection tasks or the imposition of an inspection interval which is lower than the MRB report.

When an airplane is transferred from one maintenance program to another because of a sale or lease, the principal activity is to adjust the inspection intervals to the program the airplane is being transferred to.

Escalation, windows, and proration

Escalation

Escalation is a maintenance process for interval adjustment of hard time and on-condition intervals using such reliability techniques as sampling, actuarial studies, component performance, inspector and shop findings, pilot reports, removal rates, and so on. It is based on the reliability program data primarily aimed at on-condition maintained items. Today a very limited number of hard timed items will or can be escalated.

Inspections contained in the Airworthiness Limitations Section of the MRB document cannot be increased or decreased without FAA ACO approval. If the operator adopts a formal reliability program, specific approval for individual escalation is not required.

It is extracurricular to this discussion to examine in detail the rigorous technique required to substantiate an escalation. Suffice to say that it is a structured, disciplined approach. Neither large incremental increases in inspection intervals nor large-scale abandonment of maintenance inspections is the norm.

Short-term escalation

Occasionally an operator through no fault of his own may be unable to accomplish a given inspection or task within the required interval or effect immediate repair. This frequently occurs because of schedule disruptions during the winter months. In the case of repairs, there may be difficulty obtaining the necessary materials. An operator may extend the required inspection/task beyond the required interval. This extension of time is known as short-term escalation. Under controlled conditions, short-term escalation for an individual component, engine, or aircraft may be used without affecting safety.

Short-term escalations also permit scheduling of time-controlled tasks in conjunction with other scheduled tasks when the specified intervals are different but are within close proximity of each other.

Short-term escalations must be limited to the required need and limited to a specific airplane or component. They are not used to authorize deferral of a maintenance program item beyond the first opportunity to accomplish that item without an interruption to the scheduled operation of the airplane. They may not be used to extend:

- AD note intervals
- Life-limited part intervals specified by the manufacturer or identified in airplane or engine Type Certificate data sheets
- Intervals specified in the Minimum Equipment Lists (MELs)
- Airworthiness limitations

Close monitoring of escalation is required to assure that it does not conceal unsound maintenance practices, maintenance program deficiencies, or poor management decisions. Supplemental or additional checks may be required to assure airworthiness during the escalation period. Escalations must not cause items not covered by the escalation to exceed their maintenance intervals.

Operators using a reliability program do not require prior approval before using escalation. They must, however, inform their Principal Maintenance Inspector of escalation as soon as possible after as escalation is put into effect. Short-term escalation for operators not under a reliability program must be approved in the Operations Specifications.

Short-term escalation intervals are either a percentage of existing intervals for a particular inspection, or designated in hours of service, in cycles, or in other increments. Except under certain conditions, maximum time for escalation is 500 hours time in service or its equivalent.

It should be noted that short-term escalation is now allowed for Certification Maintenance requirements.

Windows

Some programs will use a window of tolerance instead of short-term escalation for required inspections/checks, that is, a certain period + or − for early or late accomplishment of inspections is allowed. Operators authorized for short-term escalation will not be eligible for these built-in inspection tolerances. Inspection tolerances permit better scheduling. However, definite stipulations apply if they are approved:

- They must not affect the target inspection interval.
- The inspection due time or date must be plotted out and not be changed because of the use of a tolerance.

- There must be no time accumulation or sliding because of inspections performed early or late.
- The use of a window for extending an interval will require the next interval to be shorter (never longer).

Windows may not be used to piecemeal inspections. Once an inspection segment is initiated (i.e., panels opened), it should be completed before the aircraft is placed back in service.

Inspection tasks resulting from life limitations, airworthiness limitations, or calendar intervals are not authorized a tolerance. No inspection time and/or interval, mandated by an Airworthiness Directive, can be adjusted or changed, except within the context of the AD.

Proration

Proration is the process for adjusting hard time and on-condition intervals when transferring an airplane or component from one continuous airworthiness maintenance program to another. Consider the following simple example. Pump xx-xxxx-1 is hard timed at 5,000 hours under program A. The pump is transferred to maintenance program B which hard times this same pump at 9,000 hours. What is the time on the pump when it transfers to program B?

$$4{,}000/5{,}000 = X/9{,}000$$

$$X = (4{,}000/5{,}000)(9{,}000) = 7{,}200 \text{ hours}$$

Proration is also used as a business tool when purchasing used equipment, entering leases and allocating maintenance cost between two operators.

Equipment leases and interchanges

Equipment leases and interchange agreements[51] affect the structure of the maintenance program. They affect operational control of the airplane. Interchanges are a form of short-term dry lease agreement. They involve the temporary transfer of an airplane from the certificate of operator A to operator B's certificate.[52]

[51] 14 CFR 121.569, Equipment Interchange: Domestic and Flag Operations.

[52] 14 CFR 119(a)(4)(ii) requires airplanes to be specifically listed on the Operations Specifications of the operator.

The technical aspects of the lease or interchange are subject to FAA review and approval. They may be extremely complex. The configuration of the leased airplane, the program under which the airplane is being maintained, and operational control are examples of the issues that must be reviewed. Important details to be reviewed include the airplane configuration and maintenance program difference. These differences can be potentially dangerous.

This transfer of operational control requires that each operator's certificate and maintenance program be amended. Maintenance programs (lessee's or lessor's) that will be utilized must be identified. Short-term airplane leases between certificate holders may result in a requirement that the leased airplane must be operated under the approved maintenance program of the lessor. Long-term leases between certificate holders may involve transferring the airplane from the lessors maintenance program to that of the lessee.

Lease of airplanes from financial institutions is more straightforward. The lessee normally includes such airplanes directly into a maintenance program if they are new. There may, however, be complications when an airplane leased by a financial institution, which is the airplane owner, takes the airplane back from one operator and then leases it to another. This can be an extremely involved and very costly process when determining the maintenance program. It can become even more complex if the airplane was in foreign registry and is being returned to U.S. registry.

Foreign-registered airplanes

Foreign-registered airplanes may be added to an operator's certificate. This most frequently arises from aircraft short-term leases. The aircraft must be leased or chartered without crew. The country of aircraft registry must be a member country of the convention of the International Civil Aviation Organization (ICAO).

The Operations Specifications and the maintenance program will contain provisions to require that the airplane be maintained under the airworthiness rules and maintenance program of the foreign owner/operator. In order to maintain the validity of the foreign airworthiness certificate, the U.S. operator may have to perform inspections that are more extensive than those required by its FAA-approved continuous airworthiness maintenance program and/or the Federal Aviation Regulations. Individuals returning to service may need to have appropriate foreign certification.

Aircraft changes may have to be made before the U.S. operator can use a foreign aircraft. These changes may invalidate the certificate. In such cases, an exemption may be required from the foreign airworthiness authority.

Major repair and alteration

Certificate holders may approve an airplane for return to service after performing a major repair or major alteration, if the work has been done in accordance with technical data approved by the Administrator. Many times the operator relies on the manufacturer to provide data using the DER system to generate these data. Most frequently, this work is associated with structural repairs or interior modification.

Some large operators have large engineering departments with the capability to develop their own repair data. SFAR 36 allows qualified operators to obtain approval to perform major repairs and return to service using technical data that have not been approved by the Administrator.

Substance abuse programs

On November 14, 1988, the FAA issued a final rule requiring specified aviation employers and operators to initiate antidrug programs,[53] including drug testing for personnel performing specified safety-related functions. Subsequently, on October 28, 1991, the Omnibus Transportation Employee Testing Act of 1991 (the Act) was enacted. Among other things, the Act provided a statutory mandate for drug testing in the aviation industry and required specific consequences for positive drug tests.

On February 15, 1994, the Department of Transportation published final alcohol-testing rules, including a requirement that evidential breath-testing devices be used to conduct alcohol tests. Each operator must have an approved drug-testing and alcohol abuse preven-

[53] 49 USC, Chapter 451, Alcohol and Controlled Substances Testing, and 14 CFR 121, Appendix I, Drug Testing Program, and Appendix J, Alcohol Misuse Prevention Program.

tion program. It applies to all persons who perform a safety-sensitive function. This includes maintenance personnel.

Further reading

ICAO publications

Annex 13, Aircraft Accident and Incident Investigation, 8th ed., incorporating Amendments 1–9, July 1994, 37 pp.

Annex 17, Security Safeguarding International Civil Aviation against Acts of Unlawful Interference, 6th ed., March 1997, 30 pp.

Annex 18, The Safe Transport of Dangerous Goods by Air, 2d ed., incorporating Amendments 1–4, July 1989. Reprinted August 1997, 23 pp.

Accident/Incident Reporting Manual (ADREP Manual) (Doc 9156), 2d ed., 1987. Reprinted September 1999, 97 pp.

Manual of Aircraft Accident Investigation (Doc 6920). 4th ed., 1970. Reprinted September 1995, incorporating Amendments 1–11, 757 pp.

Accident Prevention Manual (Doc 9422), 1984. Reprinted March 1999, 144 pp.

Airworthiness Technical Manual (Doc 9051), 2d ed., 1987. Reprinted October 1999, incorporating Amendments 1–11, 264 pp.

Continuing Airworthiness Manual (Doc 9642), 1995, 100 pp.

Dangerous Goods Training Programme (Doc 9375), Book 1, Shippers, Cargo Agents and Operators' Cargo Acceptance Staff. 3d ed., January 1993, 124 pp.

Emergency Response Guidance for Aircraft Incidents Involving Dangerous Goods (Doc 9481), 1999–2000 edition.

Manual of Aircraft Ground De/Anti-Icing Operations (Doc 9640), 1995, 29 pp.

Technical Instructions for the Safe Transport of Dangerous Goods by Air (1999–2000 edition) (Doc 9284), 1998, 646 pp.

Transport of Dangerous Goods by Air (1999–2000 edition) (Doc 9284), 1999, 120 pp.

The Continuing Airworthiness of Aircraft in Service Codes of Airworthiness Used by Different States, Methods of Handling and Exchange of Information on Airworthiness Directives (or Their Equivalent) and Details of Systems Used in States for Reporting of Information on Faults, Defects and Malfunctions (Circ. 95), 6th ed., 1985, 182 pp.

NTSB publications

Guidance for Party Coordinators and Other Participants in the Investigation of Aircraft Accidents, NTSB Office of Aviation Safety.

FAA publications

FAA Order 8300.10, *Airworthiness Inspector's Handbook*

Volume 2:

Chapter 2, Issue SFAR 36, Authorization

Chapter 3, Evaluate Category I/II/III/IIIa, Landing Minimum Maintenance/Inspection Programs

Chapter 4, Evaluate an Operator's Deicing/Anti-Icing Program

Chapter 63, Evaluate FAR Part 121/135.411(A)(2), Company Manual/Revision

Chapter 64, Evaluate Continuous Airworthiness Maintenance Program/Revision

Chapter 65, Evaluate Continuing Analysis and Surveillance Program/Revision

Chapter 66, Approve Reliability Program

Chapter 67, Approve Contract Reliability Program

Chapter 69, Evaluate FAR Part 121, Maintenance Contractual Arrangement

Chapter 71, Evaluate FAR Part 121, Operator's Maintenance Records

Chapter 73, Evaluate FAR Part 121/135.411(A)(2), Leased Maintenance Program Authorization: U.S. Registered Aircraft

Chapter 78, Process FAR Part 121/135.411(A)(2), Operator Aircraft/Engine Utilization Report

Chapter 79, Review FAR Part 121/135.411(A)(2), Engineering Change Authorization

Chapter 80, Evaluate Short-Term Escalation Procedures

Chapter 81, Evaluate Foreign Registered Aircraft Operated by FAR Part 121/135.411(A)(2), Operators

Chapter 82, Evaluate FAR Part 121, Extended Range Operations with Two-Engine Aircraft (ETOPS)

Chapter 87, Approve Parts / Parts Pool / Parts Borrowing

Chapter 88, Prorated Time Authorizations

Chapter 89, Special Flight Permit with Continuing Authorization to Conduct Ferry Flights

Chapter 221, Conduct Evaluation of Operator/Applicant's Main Base Facility

Chapter 227, Evaluate Applicant's Refueling Procedures

Chapter 235, Introduction to Avionics

Chapter 236, Evaluate Avionics Test Equipment

Chapter 237, Evaluate Avionics Equipment Approval

Chapter 238, Evaluate Airborne Microwave Landing Systems

Chapter 239, Approve Altimeter Setting Sources

Chapter 241, Approve Area Navigation Systems

Volume 3:

Chapter 3, Conduct Ramp Inspection of Operator's Aircraft

Chapter 8, Conduct a Detailed Process/Task Inspection

Chapter 36, Monitor Continuous Airworthiness Maintenance Program/Revision

Chapter 37, Monitor Continuing Analysis and Surveillance Program/Revision
Chapter 38, Monitor Approved Reliability Program
Chapter 40, Monitor FAR Part 121, Contractual Reliability Program
Chapter 43, Monitor FAR Part 121, Extended-Range Operations with Two-Engine Aircraft (ETOPS)
Chapter 124, Issue Aircraft Condition Notice
Chapter 125, Monitor Operator during Strike/Labor Unrest/Financial Stress
Chapter 127, Monitor Operator during Mergers/Acquisitions/Bankruptcy Proceedings
Chapter 128, Process Service Difficulty Report
Chapter 129, Process Malfunction or Defect Report
Chapter 130, Review Operator's Mechanical Interruption Summary Report
Chapter 131, Inspect Operator's Main Base Facility
Chapter 135, Monitor Operator's Refueling Procedures
Chapter 142, Monitor Flight Data Recorders
Chapter 143, Monitor Cockpit Voice Recorders
Chapter 144, Inspect Avionics Test Equipment
Chapter 145, Inspect Altimeter Setting Sources
Chapter 146, Monitor Approved Avionics Software Changes

Volume 4:
Chapter 7, Powerplant Repairs
Chapter 8, Human Factors Involved in Inspection and Repair in a Heavy Maintenance Environment

Flight Standards Handbook Bulletin Airworthiness (HBAW)

HBAW 91-13, Subject: Flightcrew Erasing of Engine Indicating and Crew Alerting Systems (EICAS) Status Messages

HBAW 92-07, Maintenance of Acoustic Underwater Locator Beacons

HBAW 94-03 and HBAT 94-09, Engine Change and Verification Flights for Extended Range Operation with Two-Engine Airplanes (ETOPS)—Guidance for ETOPS Operators

*HBAW 94-05B (amended), Significant Differences between Flight Cycle and Flight Time Relationship Affecting Airplane Maintenance Programs

HBAW 94-06, Time-Limited Dispatch Approval of Engines Fitted with Full Authority Digital Engine Control Systems

*HBAW 94-08, Policy Changes to the ETOPS Configuration, Maintenance and Procedures Documents

*HBAW 95-06A (amended), Maintenance Programs for Aircraft Engines, Including Leased Engines, Used by Operators of Transport Category Aircraft

HBAW 95-10, Guidelines for Operational Approval of Global Positioning System (GPS) to Provide the Primary Means of Class II Navigation in Oceanic and Remote Areas of Operation

HBAW 96-01, Information and Guidance Pertaining to Structural Maintenance Programs for Aging Large Transport and Other Transport Category Airplanes

HBAW 96-02, Acceptance of Documentation for New Parts Received from the Original Equipment Manufacturer Located Outside the U.S.

HBAW 96-04A, FAA Policy Regarding Segmented Inspections and Built-In Inspection Tolerances (Windows) in Approved Aircraft Inspection Programs (AAIP)

HBAW 96-05C (amended), Air Carrier Operations Specifications Authorization to Make Arrangements with Other Organizations to Perform Substantial Maintenance

HBAW 96-08, Aviation Safety Inspectors (ASI) Guidance for Detecting Unapproved Parts to Accomplish PTRS Codes 3622/5622 for Air Carriers and 3668/5668 for Air Agencies

HBAW 97-02, Establishment of Aviation Safety Action Programs (ASAP)

HBAW 97-06, FAA Policy Regarding Segmented Inspections and Built-In Inspection Tolerances (WINDOWS) in Continuous Airworthiness Maintenance Programs (CAMP)

HBAW 97-07, HBAT 97-08, and HBGA 97-05, Transportation of Oxygen Generators (Chemical)

HBAW 97-08A (amended), Clarification of 14 CFR Section 145.47(b), Calibration of Inspection and Test Equipment

HBAW 97-10, Program Tracking and Reporting Subsystem (PTRS) Reporting of Suspected Unapproved Parts (SUP) Investigations and Associated Activities

HBAW 97-13B (amended), Digital Flight Data Recorder Maintenance

HBAW 97-15, Additional Procedures to the Air Carrier's Continuous Airworthiness Maintenance Program that Ensures Covers Are Removed from Pitot-Static Ports Following Cleaning and Maintenance

HBAW 98-01, Air Carrier and Maintenance Provider Contracts

HBAW 98-02, Evaluation of 14 CFR Parts 121/135, Air Carriers' Maintenance Record Keeping System

HBAW 98-03A (amended), New Entrant Air Carrier's Development of Acceptable Maintenance Program, Time Intervals, and Maintenance Process

HBAW 98-10, Air Carrier Control and Handling of Aircraft Components or Consumable Materials That Contain Hazardous Materials

HBAW 98-15A (amended), New Guidance for Operations Specifications (OpSpecs) Paragraph D085 "Aircraft Listing"

HBAW 98-17A (amended), Guidance for Evaluating Requests for Approval of Procedures Found in the NESHAP Advisory Circular

HBAW 98-21 and HBAT 98-36, Monitoring Operators during Periods of Growth or Major Change

HBAW 99-10, Air Agency Control and Handling of Aircraft Components or Consumable Materials that Contain Hazardous Materials

HBAW 99-19 and HBAW 99-16, 14 CFR Parts 121 and 135, Air Carrier Safety Departments, Programs, and the Director of Safety

Flight Standards Information Bulletin for Airworthiness (FSAW)

FSAW 94-25B (extended), Air Carrier Strobe Light Maintenance Programs

FSAW 94-45A (extended), Electronic Signatures and Exemptions from the Requirements of FAR Section 43.9

FSAW 94-47 (extended), Flightcrew Notification of Prior Maintenance

FSAW 95-02 and FSAT 95-03, List of Minimum Instruments to Ferry Large Turbojet Airplanes

FSAW 95-03, Seat Back Break-over

FSAW 96-06A (amended), Damaged and Suspected Aircraft Parts Entering the Surplus Sales Market

FSAW 96-11 and FSAT 96-06, Civil Aviation Security Additional Guidance on Hazardous Material (HAZMAT) Programs

FSAW 96-13, FSAT 96-07, and FSGA 96-04, Transportation of Oxygen Generators (Chemical)

FSAW 97-08A (amended), Procedures to Ensure Covers Are Removed from Static Ports Following Cleaning and Maintenance

FSAW 97-10, Cold Weather Servicing of Aircraft Nose Landing Gear Struts

FSAW 97-18, Manufacturers' Service Documents

FSAW 97-22, Important Safety Information Advisory on Dilutions of UCAR Aircraft Deicing/Anti-Icing Fluid ULTRA+

FSAW 98-09, Availability, Retrieval and Use of Airworthiness Technical Data Obtained via Internet Providers or Other Electronic Media

FAA advisory circulars

AC 00-46D, Aviation Safety Reporting Program

AC 20-35C, Tiedown Sense

AC 20-43C, Aircraft Fuel Control

AC 20-68B, Recommended Radiation Safety Precautions for Ground Operation of Airborne Weather Radar

AC 20-114, Manufacturers' Service Documents

AC 20-123, Avoiding or Minimizing Encounters with Aircraft Equipped with Depleted Uranium Balance Weights during Accident Investigations

AC 20-125, Water in Aviation Fuels

AC 21-29B, Detecting and Reporting Suspected Unapproved Parts

AC 43-4A, Corrosion Control for Aircraft

AC 43-15, Recommended Guidelines for Instrument Shops

AC 43.9-1E, Instructions for Completion of FAA Form 337 (OMB No. 2120-0020), Major Repair and Alteration (Airframe, Powerplant, Propeller, or Appliance)

AC 90-45A, Approval of Area Navigation Systems for Use in the U.S. National Airspace System

AC 91-13C, Cold Weather Operation of Aircraft

AC 91-44A, Operational and Maintenance Practices for Emergency Locator Transmitters and Receivers

AC 91-51A, Effect of Icing on Aircraft Control and Airplane Deice and Anti-Ice Systems

AC 91-56A, Continuing Structural Integrity Program for Large Transport Category Airplanes

AC 120-16C, Continuous Airworthiness Maintenance Programs

AC 120-17A, Maintenance Control by Reliability Methods

AC 120-30A, Reporting Requirements of Air Carriers, Commercial Operators, Travel Clubs, and Air Taxi Operators of Large and Small Aircraft

AC 120-36A, Air Carrier Dispersal Planning Program

AC 120-38, Transport Category Airplanes Cabin Ozone Concentrations

AC 120-39, Hazards of Waste Water Ice Accumulation Separating from Aircraft in Flight

AC 120-42A, Extended Range Operation with Two Engine Airplanes (ETOPS)

AC 120-59, Air Carrier Internal Evaluation Programs

AC 120-60, Ground Deicing and Anti-Icing Program

AC 120-66, Aviation Safety Action Programs (ASAP)

AC 120-67, Criteria for Operational Approval of Auto Flight Guidance Systems

AC 120-69, Use of CD-ROM Systems

AC 121-1A, Standard Operations Specifications—Aircraft Maintenance Handbook

AC 129-4, Maintenance Programs for U.S. Registered Aircraft under FAR Part 129

AC 120-28D, Criteria for Approval of Category III, Weather Minima for Takeoff, Landing, and Rollout

AC 120-29, Criteria for Approving Category I and Category II, Landing Minima for FAR 121 Operators

Hazardous materials

49 CFR 107, Hazardous Materials Program Procedures

49 CFR 171, General Information, Regulations, and Definitions

49 CFR 172, Hazardous Materials Table, Special Provisions, Hazardous Materials Communications, Emergency Response Information, and Training Requirements

49 CFR 173, Shippers—General Requirements for Shipments and Packaging

49 CFR 175, Carriage by Aircraft

AC 121-21B, Information Guide for Training Programs and Manual Requirements in the Air Transportation of Hazardous Materials

AC 121-27, Information Guide for Training Programs and Manual Requirements in the Air Transportation of Hazardous Materials

AC 121-28, Preparation and Loading of Magnetic Materials for Air Transportation

13

The Maintenance Organization

Required management personnel

Certificate holders must have sufficient qualified management and technical personnel[1] to ensure the highest degree of safety in its operations. A minimum of at least five full-time management positions (or equivalent) are required and listed in the carrier's Operations Specification. The experience requirements are listed in the regulations,[2] although in some instances deviations to these conditions may be allowed. The positions follow.

Director of safety

There is no specific qualifications listed in the regulations for this position. 14 CFR 119.65(d)(1) and (d) (2) state that all required management personnel must "Be qualified through training, experience, and expertise.... To the extent of their responsibilities, have a full understanding of...aviation safety standards and safe operating practices."

Director of operations

- Hold an airline transport pilot certificate.
- Have at least 3 years of supervisory or managerial experience within the last 6 years in a position that exercised operational control over any operations conducted with large airplanes under Part 121 or Part 135 or, if the certificate holder uses only small airplanes in its

[1] 14 CFR 119.65.

[2] 14 CFR 119.65.

operations, the experience may be obtained in large or small airplanes.

- For the first time ever, have at least 3 years of experience within the past 6 years as pilot in command of a large airplane operated under Part 121 or Part 135, if the certificate holder operates large airplanes. If the certificate holder uses only small airplanes in its operation, the experience may be obtained in either large or small airplanes.
- In the case of a person with previous experience as a director of operations, have at least 3 years of experience as pilot in command of a large airplane operated under Part 121 or Part 135, if the certificate holder operates large airplanes. If the certificate holder uses only small airplanes in its operation, the experience may be obtained in either large or small airplanes.

Chief pilot

- Hold an airline transport pilot certificate with appropriate ratings for at least one of the airplanes used in the certificate holder's operation.
- If becoming a chief pilot for the first time ever, have at least 3 years of experience within the past 6 years as pilot in command of a large airplane operated under Part 121 or Part 135, if the certificate holder operates large airplanes. If the certificate holder uses only small airplanes in its operation, the experience may be obtained in either large or small airplanes.
- In the case of a person with previous experience as a chief pilot, have at least 3 years of experience as pilot in command of a large airplane operated under Part 121 or Part 135, if the certificate holder operates large airplanes. If the certificate holder uses only small airplanes in its operation, the experience may be obtained in either large or small airplanes.

Director of maintenance

- Hold a mechanic certificate with airframe and power plant ratings.
- Have 1 year of experience in a position responsible for returning airplanes to service.

- Have at least 1 year of experience in a supervisory capacity, maintaining the same category and class of airplane as the certificate holder uses.
- Have 3 years of experience within the past 6 years in one or a combination of the following:
  - Maintaining large airplanes with 10 or more passenger seats, including at the time of appointment as director of maintenance, experience in maintaining the same category and class of airplane as the certificate holder uses.
  - Repairing airplanes in a certificated airframe repair station that is rated to maintain airplanes in the same category and class of airplane as the certificate holder uses.

Chief inspector

- Hold a mechanic certificate with both airframe and power plant ratings, and have held these ratings for at least 3 years.
- Have at least 3 years of maintenance experience on different types of large airplanes with 10 or more passenger seats with an air carrier or certificated repair station, 1 year of which must have been as maintenance inspector.
- Have at least 1 year of experience in a supervisory capacity, maintaining the same category and class of aircraft as the certificate holder uses.

Management personnel qualifications

All the required individuals and any other individuals in a position to exercise control over operations conducted under an operating certificate must:

- Be qualified through training, experience, and expertise.
- To the extent of their responsibilities, have a full understanding of the following materials with respect to the certificate holder's operation:
  - Aviation safety standards and safe operating practices
  - Federal Aviation Regulations
  - The certificate holder's operations specifications
  - The manual required by 14 CFR 121.133
- Discharge their duties to meet applicable legal requirements and to maintain safe operations.

Persons authorized to perform maintenance

The regulations are quite simple in designating who may perform maintenance:

- The holder of a mechanic certificate with appropriate rating
- The holder of a repairman's certificate but only for that work which is applicable to his repairman rating and work which was accomplished by the worker
- A person working under the supervision of the holder of a mechanic or repairman certificate
- The holder of a repair station certificate for work which the repair station is authorized[3]
- The holder of an air carrier operating certificate under 14 CFR 121 through the limitations of the approved maintenance program and as listed in the Operations Specifications

A manufacturer may perform maintenance within the following limitations:

- The device maintained was manufactured under a Type or Production Certificate or a Technical Standard Order.

The holder of a pilot certificate may perform preventive maintenance owned or operated by that pilot. However, the airplane may not be used in operations conducted under 14 CFR 121. The authorization realistically extends to only two persons, that is, a certificated mechanic or repairman. Repair stations and air carriers are agencies. Their agent for performing maintenance is a mechanic or a repairman.

Noncertificated persons (with certain restrictions) working under the supervision of a holder of a mechanic or repairman certificate may also perform maintenance. There is increasing concern in the community as to the number of noncertificated persons working on airplanes under mechanic/repairman supervision. The number is getting larger and larger. The consequences are that the ability to effectively oversee large groups of people is in question. The problem is attributable in part to an increasing shortage of certificated people.

[3]In the case of U.S.-registered airplanes only a 14 CFR 145 certificated repair station. Thus, for example, work done by say Lufthansa on a Burning Stump U.S.-registered aircraft must be done under the terms of Lufthansa's United States Foreign Repair Station Certificate.

Mechanic/repairman's certification

The licensing of mechanics[4] and repairman, addressed in 14 CFR 65, is summarized in Table 13-1.

Essentially the difference between a mechanic and a repairman is that a mechanic has severe restrictions on the certificate. Repairmen are certificated on their experience and the recommendation of their employers. They may only work on specific items at a specific location for a specific employer.

It is possible to obtain certification on the basis of experience. However, mechanics are frequently graduates of the approved schools defined in 14 CFR 147, Aviation Maintenance Technician Schools. The 1-year curriculum requires at least the following minimum hours of instruction:

- Airframe—1,150 hours (400 general plus 750 airframe)
- Power plant—1,150 hours (400 general plus 750 power plant)
- Combined airframe and power plant—1,900 hours (400 general plus 750 airframe and 750 power plant)

The curriculum, divided into three categories, covers the topics shown in Table 13-2.

Testing

Applicants for a mechanic certificate must complete a written, oral, and practical application for each of the general, airframe, and power plant curricula.

What do mechanics do?

There is a perception that mechanics have no part in the airworthiness process other than to do the work and keep quiet. Nothing is further from the truth. Certificated mechanics are the final gate in the airworthiness chain. Notice that they are the only individuals authorized to return to service.

[4]Currently called AMTs. This euphemism was developed during the 1990s. Its use has been justified to improve the image of mechanics. It was felt that the word mechanic had too many negative connotations attached to it. I oppose this term. It is patronizing. It panders the very image it intends to improve. Mechanics are extremely perceptive. They see right through this nonsensical attempt at political correctness. Also I was not aware that mechanic is a dirty word.

TABLE 13-1 Mechanic/Repairman Certification.

Summary Mechanic/Repairman Certification—14 CFR 65	
	Mechanic
Eligibility	• Be at least 18 years of age • Be able to read, write, speak, and understand the English language, with some exceptions • Pass all of the prescribed tests within 24 months
Ratings	• Airframe. • Power plant.
Knowledge	• Pass a written test covering the construction and maintenance of aircraft appropriate to the rating sought, the regulations • Pass each section of the test before applying for the oral and practical tests prescribed.
Experience	• Graduate from a certificated aviation maintenance technician school or • At least 18 months of practical experience with the procedures, practices, materials, tools, machine tools, and equipment generally used in constructing, maintaining, or altering airframes, or power plants appropriate to the rating sought or • At least 30 months of practical experience concurrently performing the duties appropriate to both the airframe and power plant ratings.
Skill	• Must pass an oral and practical test on the rating(s) sought.
Privileges and limitations	• May perform or supervise the maintenance, preventive maintenance or alteration of an aircraft or appliance, or a part thereof, for which rated (excluding major repairs/alterations of, propellers, and any repair to/alteration of, instruments) • May not supervise maintenance, preventive maintenance, or alteration of, or approve and return to service, any aircraft or appliance, or part thereof, for which rated unless has satisfactorily performed the work concerned at an earlier date.

TABLE 13-1 Mechanic/Repairman Certification. (*Continued*)

Summary Mechanic/Repairman Certification—14 CFR 65	
Mechanic	
	• If not, show ability to do it by performing under the direct supervision of an appropriately rated mechanic/repairman, who has had previous experience in the specific operation concerned. • May not exercise the privileges of certificate and rating unless understands the current instructions of the manufacturer, and the maintenance manuals, for the specific operation concerned.
Recent experience	• May not exercise the privileges of his certificate and rating unless, within the preceding 24 months: • Administrator has found that he is able to do that work or • He has, for at least 6 months — (1) Served as a mechanic under his certificate and rating; (2) Technically supervised other mechanics; (3) Supervised, in an executive capacity, the maintenance or alteration of aircraft
Repairman Certification Summary	
Eligibility	• Be at least 18 years of age; • Be specially qualified to perform maintenance on aircraft or components thereof, appropriate to the job for which employed; • Be employed for a specific job requiring those special qualifications by a certificated repair station, or by a certificated commercial operator or certificated air carrier; • Be recommended for certification by his employer; • Have either — (1) At least 18 months of practical experience in the specific job for which employed, or (2) Completed formal training • Be able to read, write, speak, and understand the English language, with some exceptions.

TABLE 13-1 Mechanic/Repairman Certification. (*Continued*)

Summary Mechanic/Repairman Certification—14 CFR 65	
	Repairman Certification Summary
Privileges and limitations	• Perform or supervise the maintenance, preventive maintenance, or alteration of aircraft or aircraft components appropriate to the job for which employed and certificated, but only in connection with duties for the certificate holder by whom employed and recommended. • May not perform or supervise duties under the repairman certificate unless understands the current instructions of the certificate holder by whom employed and the manufacturer's instructions for continued airworthiness relating to the specific operations concerned.

They are the primary determinant of what needs to be fixed. They are the troubleshooters. They assess the meaning and intent not only of the regulations, but also of the myriad of (and all too frequently confusing and incorrect) technical specifications, instructions, and performance standards. And they effect repairs and replacement on the basis of this knowledge and their mechanical skills every day.

What are mechanics like?

Mechanics are guarded and somewhat distrustful of "outsiders." They are extremely pragmatic and insightful, if somewhat uncommunicative. They are very well educated. They continue their education after mechanic school taking additional technical, business, and management courses. Many have undergraduate degrees in a number of disciplines including engineering. A surprising number have advanced degrees. They are self-starters and independent thinkers. Taylor and Christensen, in their book, describe the personality of mechanics well.

> *AMT job descriptions are one thing. The people who actually meet these requirements tend to share certain other characteristics as well. General observations of what AMTs are actually like on the job have been circulating for many years....*

Table 13-2 Mechanic Curriculum.

Airframe and Power Plant Mechanic Curriculum				
	Airframe		Power Plant	
General	Structures	Systems	Theory and Maintenance	Systems
Basic Electricity	Wood Structures	Landing Gear	Reciprocating Engines	Instruments
Aircraft Drawings	Aircraft Covering	Hydraulic	Turbine Engines	Fire Protection
Weight and Balance	Aircraft Finishes	Pneumatic	Engine Inspection	Electrical
Fluid Lines and Fittings	Sheet Metal and Nonmetallic Structure	Cabin Atmosphere Control		Lubrication
Materials and Processes	Welding	Aircraft Instruments		Ignition and Starting
Ground Operation and Servicing	Assembly and Rigging	Communication and Navigation		Fuel Metering
Cleaning and Corrosion Control	Airframe Inspection	Fuel		Fuel
Mathematics		Electrical		Induction and Airflow
Maintenance Forms and Records		Position and Warning		Cooling
Basic Physics		Ice and Rain Protection		Exhaust and Reverser
Maintenance Publications		Fire Protection		Propellers
Mechanic Privileges and Limitations				Unducted Fans
				Auxiliary Power Unit

> *The list* [Table 13-3] *says a lot about AMTs and their general attitudes toward communications. AMTs are strong and silent types, loners who seldom communicate at all, much less in group discussions. And AMTs have themselves come to believe that they cannot communicate very well for the simple reason that their work system does not expect them to be able to communicate very well....*
>
> *The fact is that in their professional life they are normally in a "one down" position. Management, pilots and engineers all look down on them—or at least that's the way many AMTs see it. Working as equals with these people brings out the best in the mechanic and in the system as a whole.*

Mechanics love to be irreverent.[5] They may at times be uncommunicative, but they are notorious for their banter and allegory. They delight in being mischievous. They will deceptively test the credentials of anyone who is not one of them. They are self-deprecating particularly to deprecating behavior directed toward them. But don't be misled. Mechanics are not irresponsible buccaneers whose hat

TABLE 13-3 What characterizes the AMT.

What Characterizes the AMT?
Commitment to excellence
Willingness to put in effort and hours
Integrity
Distrust of words
Dependability
The tendency to be a loner
Modesty (no desire to be in the spotlight)
Doesn't like to ask for help
Tends to be self-sufficient
Doesn't share thoughts too frequently

Source: Taylor, James C., and T. J. Christensen, *Airline Maintenance Resource Management*, Society of Automotive Engineers, Inc., Warrendale, PA: ISBN 0-7680-0231-1. Cited in Richardson Management Associates, Ltd., 2054 Sherbrooke St. West, Montreal, Quebec H3H 1G5, Canada; cited in K. D. Forth, "Squeaky Wheel Gets Grease and Training," *Aviation Equipment Maintenance* (August 1993):56.

[5]Especially line maintenance (gate) mechanics.

size, collar size, belt size, and IQ are all the same number—48. Their irreverent mischievous persona also keeps them from worrying too much about what they do every time they sign off an airplane. If you don't have a mechanic's certificate in your pocket, you probably have never thought about the following. Consider that a mechanic, who works on an airplane and signs it off, must subsequently watch the airplane he declared to be safe go down the runway with 330 people on board. They always worry that they missed something.

Maintenance department organization

Each airline is organized differently with various titles and structures. Regardless of the particular organizational structure, certain elements must be present within the organization to meet regulatory requirements and elements of the maintenance program.

A senior management official usually called the vice president of maintenance and engineering who reports directly to the president of the airline normally heads the organization (Figure 13-1). The following are the common elements of any air carrier maintenance organization.

Staff

The organization is supported by the normal organizations found in any business venture. These typically include such groups as accounting, budgets, cost control, personnel administration, and several administrative groups.

Technical services

Technical services (Figure 13-2) provide engineering support to the maintenance organizations. They are the communications interface with regulatory agencies, manufacturers, and industry associations. Their principal product is the design, plans, and instructions necessary to conduct alterations and repairs. The group is either organized by fleet types or specific discipline, that is, flight controls, fuel, structures, and so forth. Personnel usually have an engineering degree or are experienced technicians. Typical functions include the following.

Standards engineering

This organization defines the processes and technical standards used in maintenance policies, practices, and procedures. They ensure that

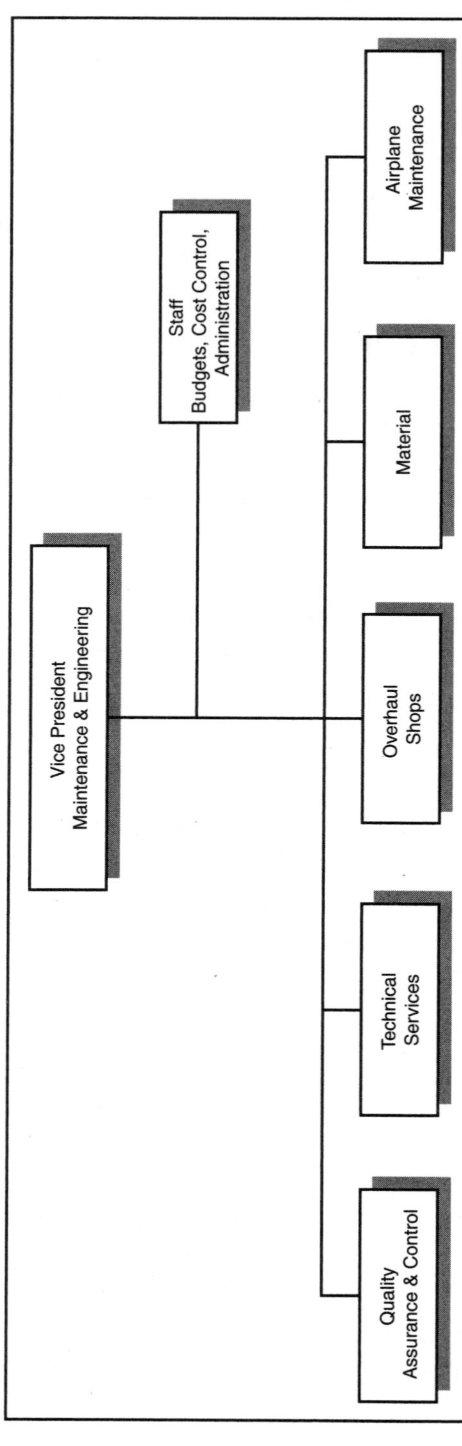

Figure 13-1 *Air carrier organization—VP Maintenance.*

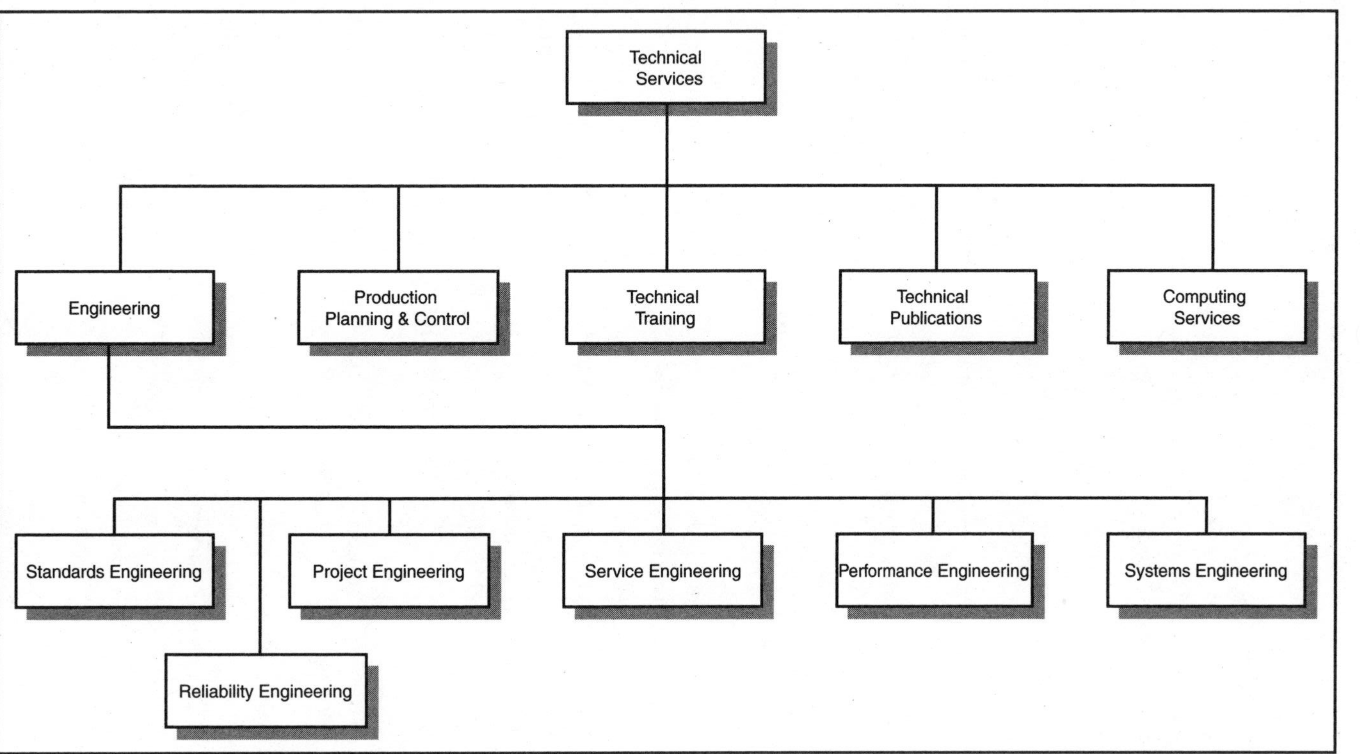

Figure 13-2 *Air carrier organization—technical services.*

technical manuals and programs are current, accurate, and in compliance with relevant regulations.

Project engineering

Project engineering is involved in nonroutine projects. It specifies the inspection and production actions that must be accomplished for each project. Work includes:

- Determine applicability, compliance requirements, time limits, and methods of accomplishment for airworthiness directives
- Oversee, evaluate, plan, budget, and incorporate manufacturer's service bulletins and other items deemed cost effective
- Develop plans, budgets, and engineering orders for airplane modifications such as new interiors, post new airplane delivery projects, equipment, systems alterations, etc.
- Obtain Supplemental Type Certificates

Service engineering

Service engineering directly supports day-to-day maintenance operations. Typical activities include:

- Give engineering assistance to the maintenance control center, hangars, and shops to correct in-service problems
- Investigate repetitive airplane discrepancies to isolate causes and develop remedial action
- Interpret technical data and resolve differences
- Design special repairs for items not covered in manufacturers manuals
- Work with industry associations and manufacturers representatives to solve airplane service difficulties

Performance engineering

Performance engineering is involved in a number of unique disciplines. The performance engineering function is closely related to flight operations. Their activities include:

- Establish and maintain the weight and balance program
- Evaluate and control the engine condition–monitoring program
- Audit and evaluate airplane and power plant performance and initiate actions to reduce fuel burn and airplane drag

- Prepare takeoff, en route, and landing performance analysis for use by flight crews and dispatcher when determining performance-limiting weights
- Prepare airplane performance materials and fuel burn tables for use by flight crew and dispatch
- Control flight-planning computer programs for use by flight dispatch/flight following
- Prepare and control the Minimum Equipment List
- Assist in new airplane evaluation and selection

Reliability engineering

Reliability engineering administers the reliability program. Their reports lead to changes in the maintenance program. They investigate items exceeding defined performance-level items and develop corrective action to clear them.

Production and planning and control

This organization prepares and controls the maintenance plan. They work closely with marketing, flight schedule planning, finance, and flight operations. Planning takes advantage of periods with low airplane utilization to remove airplanes from the revenue schedule to complete maintenance work, especially discretionary work.

- Using a demand curve derived from planned revenue flying, they forecast:
  - Line maintenance requirements to meet seasonal demands
  - Scheduled check requirements
  - Parts and material usage
- They prepare and administer the airplane plot[6] working with line maintenance and flight dispatch to fit the revenue schedule.
  - It is the focal point for coordinating production work.
  - They construct and administer the maintenance program identifying scheduled and unscheduled maintenance work. Use the maintenance-planning document as a tool for standards to establish the time to do the task, frequency needed for the task, and materials required.

[6]Airplane plots are a dynamic flowchart which tracks each airplane as it moves from station to station when it is committed to the revenue schedule. They are continually being revised because of aircraft damage, flight cancellations, weather diversions, etc.

  - They establish the maintenance schedule plan that includes the work to be done, resources such as facilities, material, and people; time to do tasks.
  - They prepare task cards and other work documents for both mandatory and discretionary work.
  - They package scheduled checks including mandatory and discretionary items.
  - They monitor production performance comparing it to the plan.
- They prepare facilities, ground support, spares, and labor forecasts to line and base maintenance.

Technical training

Technical training prepares curriculum and materials, and administers various training programs. Several of these are mandated by various regulatory agencies.

Maintenance training

Includes numerous course types:

- *Indoctrination instruction.* This covers maintenance operations, policies, procedures, manuals, forms, worksheets, and so on. This can include appropriate instruction to third-party maintenance providers.
- *Airplane model type courses.* These include general airplane instruction; mechanical, electrical, and electronic courses; and structural repair instruction.
- *Recurrent training.*
- *Special methods and processes maintenance instruction.* This can include courses in new industrial processes or methods; special maintenance instruction, for example, ETOPS maintenance, or special instruction for accomplishing AD inspections, accomplishment, and so on.
- *Specialist qualification and requalification.* This addresses extra training for personnel certification for unique critical or hazardous skills or processes. Typical examples include airplane taxi, engine run-up, fuel tank entry, composite repair, and welding. Nondestructive test process qualification for dye penetrant, X-ray, magnetic particle, and eddy current inspection are included.
- *Inspector qualification instruction.*

Security training

Operators must implement a security program.[7] This includes procedures prohibiting unauthorized access to the airplane and security inspection of airplanes if airplanes are left unattended before they are placed in service. Instruction is given to appropriate maintenance personnel regarding pertinent elements of the security program.

Hazardous materials training

Maintenance operations handle and use many hazardous materials. The regulations require that employees who load, unload, or handle hazardous materials must be given instruction[8] to enable them to recognize and identify hazardous materials and handle them.

Technical publications

The regulations require a large body of materials be prepared in a required manual. The Technical Publications Group has overall responsibility to prepare and publish the required manual materials and distribute them to appropriate personnel within the organization. They assure the integrity and conformity of the publications. They also administer a technical library.

Computing services

Computing services meet the needs for computational capability, Internet/intranet services, and digital records systems, publishing, management and planning programs, establishment of networks, and other computational services.

Quality assurance and control

The functions of the quality group (Figure 13-3) include:

- Administer the Required Inspection Items (RII) including inspector qualification.
- Oversee quality and conformity of contracted maintenance.
- Support the material department in assuring certificated spares and approved materials are used.
- Administer the reliability program.

[7] 14 CFR 108, Airplane Operator Security.

[8] 49 CFR 172, Subpart H, defines training requirements.

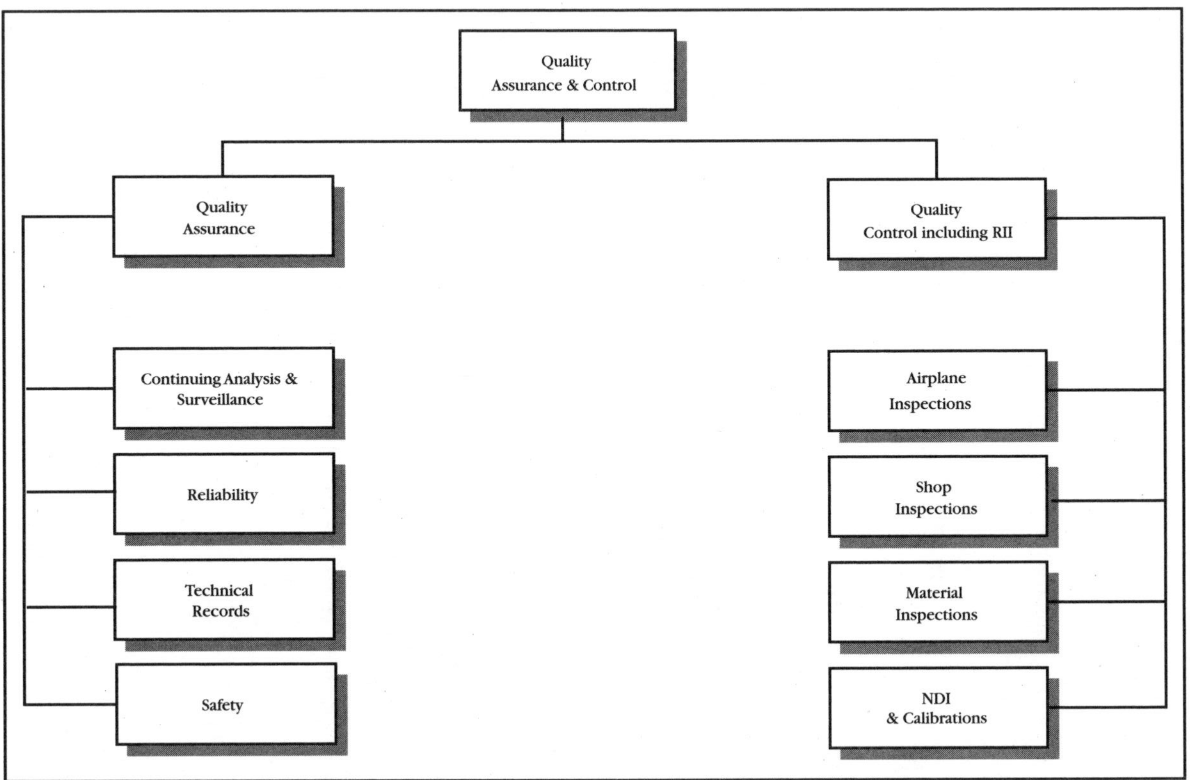

Figure 13-3 *Air carrier organization—quality assurance.*

- Manage the continuing analysis and surveillance program.
- Maintain the technical records system.

Material

Material controls (Figure 13-4) and distributes the spares and materials inventory. Their work includes:

- Work with the quality organization to make sure that parts and materials used are certified. They continuously monitor certification status of all spares and materials suppliers.
- Assure spare parts interchangeability.
- Manage software configuration control and contamination control, which is a large and growing spares issue.
- Manage parts borrowing and pooling agreements.
- Provide suitable storage facilities, including proper environment and protection.
- Check spare parts and materials. They deteriorate with time and must be discarded, inspected, or overhauled.
- Manage the unapproved spare parts monitoring program.
- Control shelf life for items that have established cure dates or shelf-life limits.

Overhaul shops

The shops (Figure 13-5) provide off-airplane and overhaul of parts:

- The engine shop does refurbishing, overhaul, and repair of power plant and APUs, including accessories. They do power plant buildup.
- Component shops accomplish electrical, mechanical, and avionic component repair.
- Support shops include ground-support equipment and machine, plating, cleaning, paint, sheet metal, and composite shops.

Airplane maintenance

This organization (Figure 13-6) is divided into hangar and line maintenance.

Hangar maintenance

This is frequently referred to as base maintenance. Hangar maintenance performs the majority of the scheduled inspections. This

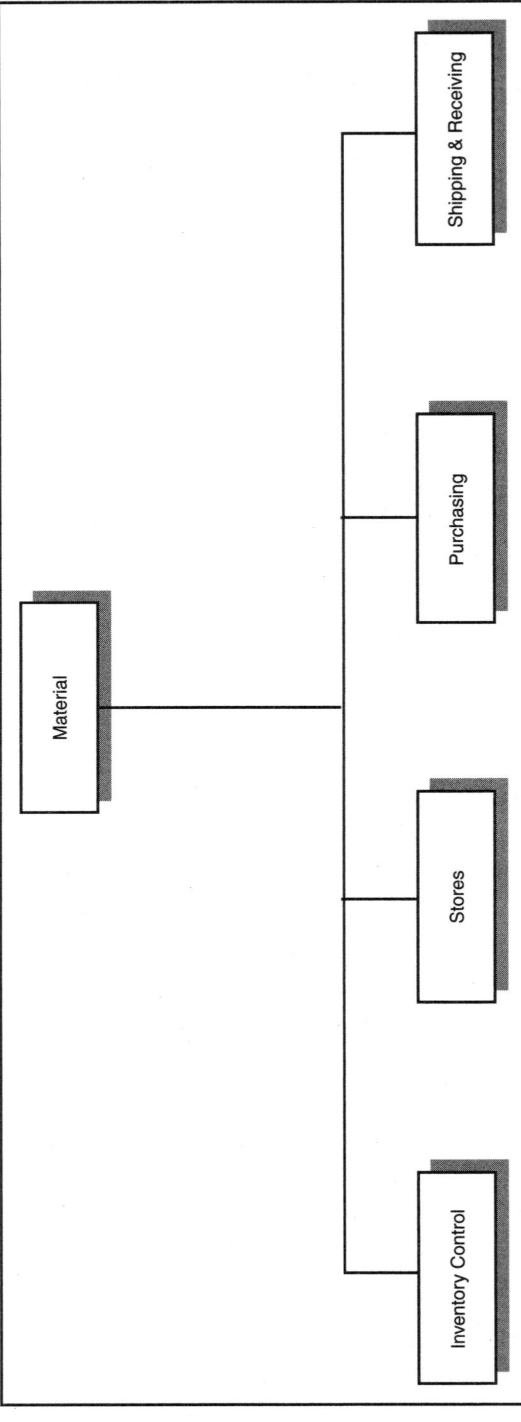

Figure 13-4 *Air carrier organization—material.*

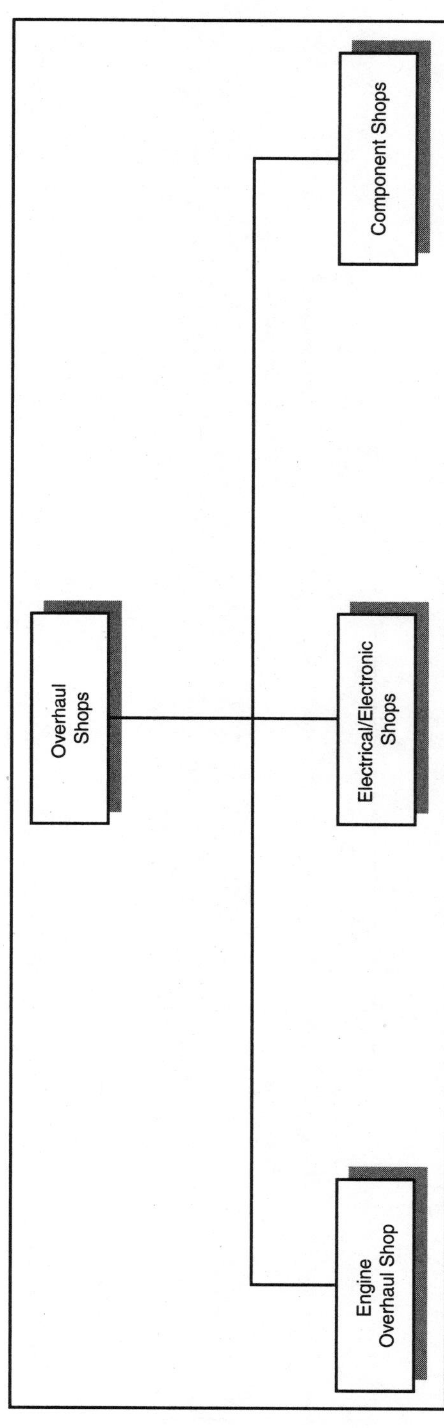

Figure 13-5 *Air carrier organization—overhaul shops.*

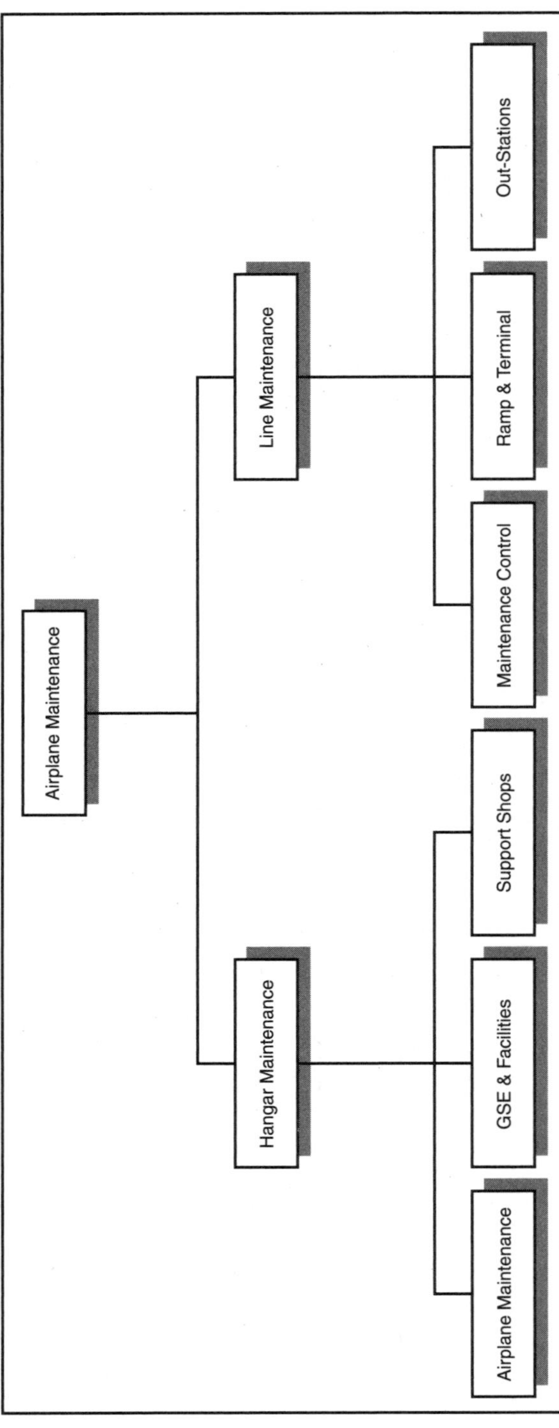

Figure 13-6 *Air carrier organization—airplane maintenance.*

group usually accomplishes the heavy checks. They accomplish modifications and repairs, clear deferred items, accomplish engineering orders, and incorporate service bulletins and AD notes.

Line maintenance

The line maintenance organization supports the airplane at individual stations in the route system while the airplanes are committed to the revenue schedule.

Maintenance control centers

Maintenance control centers manage maintenance work on airplanes committed to the revenue schedule. They work closely with flight dispatch, engineering line maintenance, and material to support the schedule and individual line stations. They coordinate deferred maintenance activities including MEL deferrals. Maintenance control extends the reach of individual gate mechanics providing additional technical expertise to them when troubleshooting and effecting repairs.

Further reading

Taylor, James C., and T. Christensen, *Airline Maintenance Resource Management,* Society of Automotive Engineers, Inc., Weirsdale, PA, 1998.

ICAO Annex 1, Personnel Licensing

ICAO Annex 6, Operation of Aircraft

FAA Orders

8300.10, *Airworthiness Inspector's Handbook*

FAA handbook

FAA-H-8083-9, *Aviation Instructor's Handbook*

Volume 2:

- Chapter 22, Certificate Airframe and/or Powerplant Mechanic/Added Rating
- Chapter 23, Certificate Foreign Applicants Located Outside the United States for Mechanic Certificates/Ratings
- Chapter 24, Certificate Repairman/Added Privileges
- Chapter 62, Evaluate FAR Part 121, Management Personnel Qualifications
- Chapter 186, Certificate FAR Part 147, Aviation Maintenance Technician School
- Chapter 187, Evaluate FAR Part 147, Aviation Maintenance Technician School's Curriculum/Revision and Instructor Qualifications
- Chapter 188, Evaluate FAR Part 147, Aviation Maintenance Technician School Facilities, Equipment, Materials, Tools, and Records

Chapter 202, Designate/Renew Designated Mechanic Examiner (DME) or Designated Parachute Rigger Examiner (DPRE)

Volume 3:

Chapter 17, Monitor Certificated Airframe and/or Powerplant Mechanic, Repairman, Parachute Rigger, and Inspection Authorization Holder

Chapter 18, Conduct a Reexamination Test of a Mechanic or an Inspection Authorization under Section 609 of the FAA Act of 1958, as Amended

Chapter 105, Inspect FAR Part 147, Aviation Maintenance Technician School

Flight Standards Handbook Bulletin (HBAW)

HBAW 95-05, Rescind Endorsement of Foreign Mechanic Applicants

HBAW 97-16, HBAT 97-13, Deviations from Part 119 for Management Personnel

Flight Standards Information Bulletin for Airworthiness (FSAW)

FSAW 94-44 and FSGA 94-11, Application Procedures for Designated Pilot Examiners (DPE), Designated Maintenance Examiners (DME), and Designated Parachute Rigger Examiners (DPRE)

FSAW 99-09, Issuance of Temporary Airman Certificate (FAA Form 8060-4) by Designated Mechanic Examiners (DME)

FAA Advisory Circulars

AC 43.13-1B, Acceptable Methods, Techniques, and Practices—Aircraft Inspection and Repair

AC 43.13-2A, Acceptable Methods, Techniques, and Practices—Aircraft Alterations

AC 65-9A, Airframe and Powerplant Mechanics General Handbook

AC 65-12A, Airframe and Powerplant Mechanics Powerplant Handbook

AC 65-15A, Airframe and Powerplant Mechanics Airframe Handbook

AC 65-2D, Airframe and Powerplant Mechanics Certification Guide

AC 65-11B, Airframe and PowerPlant Mechanics Certification Information

AC 65-24, Certification of a Repairman (General)

AC 65-25A, Aviation Maintenance Technical Awards Program

AC 65-26B, Charles Taylor "Master Mechanic" Award

AC 65-28, Aviation Mechanic General, Airframe, and Powerplant Knowledge Test Guide

AC 147-2DD, Directory of FAA Certificated Aviation Maintenance Technician Schools

AC 147-3, Certification and Operation of Aviation Maintenance Technician Schools

AC 147-4, Guidelines for Evaluation of Military Aviation Training Courses

Appendix

Definitions

Air Carrier is any citizen of the United States who undertakes directly or indirectly or by lease or any other arrangement to engage in air transportation.

Air Commerce means foreign air commerce, interstate air commerce, the transportation of mail by aircraft, the operation of aircraft within the limits of a federal airway, or the operation of aircraft that directly affects, or may endanger safety in, foreign or interstate air commerce.

Aircraft means any contrivance invented, used, or designed to navigate, or fly in, the air.

Aircraft Engine means an engine used, or intended to be used, to propel an aircraft, including a part, appurtenance, and accessory of the engine, except a propeller.

Airman means an individual:

- Who navigates aircraft when under way, in command, or as pilot, mechanic, or member of the crew.
- Who is directly in charge of inspecting, maintaining, overhauling, or repairing aircraft, aircraft engines, propellers, or appliances, except to the extent the administrator of the Federal Aviation Administration may provide otherwise for individuals employed outside the United States.
- Who serves as an aircraft dispatcher or air traffic control–tower operator.

Airport means a landing area used regularly by aircraft for receiving or discharging passengers or cargo .

Air Transportation (1) means foreign air transportation, interstate air transportation, or the transportation of mail by aircraft (Title 49 USC).

Air Transportation (2) is the carriage by aircraft of persons or property as a common carrier for compensation or hire or the carriage of mail by aircraft in commerce (FAR 119).

All-Cargo Operation means any operation for compensation or hire that is other than a passenger-carrying operation or, if passengers are carried they are only those specified in FAR 121.583(a) or 135.85.
Appliance means an instrument, equipment, apparatus, a part, an appurtenance, or an accessory used, capable of being used, or intended to be used, in operating or controlling aircraft in flight, including a parachute, communications equipment, and another mechanism installed in or attached to aircraft during flight, and not a part of an aircraft, aircraft engine, or propeller.
Cargo means property, mail, or both.
Charter Air Carrier means an air carrier holding a Certificate of Public Convenience and Necessity that authorizes it to provide charter air transportation.
Citizen of the United States means:

- An individual who is a citizen of the United States
- A partnership each of whose partners is an individual who is a citizen of the United States
- A corporation or association organized under the laws of the United States or a State, the District of Columbia, or a territory or possession of the United States, of which the president and at least two-thirds of the board of directors and other managing officers are citizens of the United States, and in which at least 75 percent of the voting interest is owned or controlled by persons that are citizens of the United States

Civil Aircraft means an aircraft except a public aircraft.
Civil Aircraft of the United States means an aircraft registered under Chapter 441 of USC Title 49.
Common Carriage The Federal Aviation Act of 1958 uses the term "common carriage" but does not define it. It is a common law term which means any operation for compensation or hire in which an operator holds itself out (by advertising or other means) as willing to furnish transportation for any member of the public who seeks that operator's services. The absence of tariffs or rate schedules, transportation only pursuant to separately negotiated contracts, or occasional refusals to transport, are not conclusive proof that the carrier is not a common carrier.

There are four elements in defining a common carrier: (1) a holding out of a willingness to (2) transport persons or property, (3) from place to place, (4) for compensation. This "holding out," which makes a person a common carrier, can be done in many ways and it does not matter how it is done.

Signs and advertising are the most direct means of holding out but are not the only ones. It may also be accomplished through the actions of agents, agencies, or salesmen who may, themselves, procure passenger traffic from the general public and collect them into groups to be carried by the operator. It is particularly important to determine if such agents or salespeople are in the business of selling transportation to the traveling public not only through the "group" approach but also by individual ticketing on known common carriers.

Holding out may be done without advertising where a reputation to carry all customers is gained. There are many means by which physical "holding out" may take place. For example, the expression of willingness to all customers with whom contact is made that the operator can and will perform the requested service is sufficient. The fact that the holding out generates little success is of no consequence. The nature and character of the operation are the important issue.

A carrier holding itself out as generally willing to carry only certain kinds of traffic is, nevertheless, a common carrier. For instance, a carrier authorized or willing only to carry planeloads of passengers, cargo, or mail on a charter basis is a common carrier, if it so holds itself out. This is, in fact, the basic business of supplemental air carriers.

A carrier flying charters for only one organization may be a common carrier if membership in the organization and participation in the flights are, in effect, open to a significant segment of the public. Similarly, a carrier which flies planeload charters for a common carrier, carrying the latter's traffic, engages in common carriage itself.

Occasionally, offers of free transportation have been made to the general public by hotels, casinos, etc. In such cases, nominal charges have been made which, according to the operators, bear the expense of gifts and gratuities. However, the operators maintain that the transportation is free. The courts have held that such operations are common carriage, based on the fact that the passengers are drawn from the general public and the nominal charge constitutes compensation.

Commuter Operation means any scheduled operation conducted by any person operating one of the following types of aircraft with a frequency of operations of at least five round trips per week or at least one route between two or more points according to the published flight schedules:

- Airplanes, other than turbojet powered airplanes, having a maximum passenger seat configuration of 9 seats or less, excluding each crewmember seat, and a maximum payload capacity of 7,500 pounds or less
- Rotorcraft

Conditional Sales Contract means a contract

- For the sale of an aircraft, aircraft engine, propeller, appliance, or spare part, under which the buyer takes possession of the property but title to the property vests in the buyer at a later time on:
  - Paying any part of the purchase price
  - Performing another condition
  - The happening of a contingency
- To bail or lease an aircraft, aircraft engine, propeller, appliance, or spare part, under which the bailee or lessee:
  - Agrees to pay an amount substantially equal to the value of the property
  - Is to become, or has the option of becoming, the owner of the property on complying with the contract

Conveyance means an instrument, including a conditional sales contract, affecting title to, or an interest in, property.

Direct Air Carrier means a person who provides or offers to provide air transportation and who has control over the operational functions performed in providing that transportation.

Domestic Operation means any scheduled operation conducted by any person operating any airplane described below at the locations described:

- Airplanes:
  - Turbojet-powered airplanes
  - Airplanes having a passenger seat configuration of more than 9 passenger seats, excluding each crewmember seat
  - Airplanes having a payload capacity of more than 7,500 pounds

- Locations:
  - Between any points within the 48 contiguous States of the United States or the District of Columbia, or
  - Operations solely within the 48 contiguous States of the United States or the District of Columbia, or

- Operations entirely within any State, territory, or possession of the United States, or
- When specifically authorized by the (FAA) administrator, operations between any point within the 48 contiguous States of the United States or the District of Columbia and any specifically authorized point located outside the 48 contiguous States of the United States or the District of Columbia

Failure A loss of function, or a malfunction, of a system or a part thereof.

Failure Condition The effect on the airplane and its occupants, both direct and consequential, caused or contributed to by one or more failures, considering relevant adverse operational or environmental conditions. Failure conditions may be classified according to their severities.

Minor Failure Failure conditions that would not significantly reduce airplane safety, and that involve crew actions that are well within their capabilities. Minor failure conditions may include, for example, a slight reduction in safety margins or functional capabilities, a slight increase in crew workload, such as routine flight plan changes, or some inconvenience to occupants.

Major Failure Failure conditions that would reduce the capability of the airplane or the ability of the crew to cope with adverse operating conditions to the extent that there would be, for example, a significant reduction in safety margins or functional capabilities, a significant increase in crew workload or in conditions impairing crew efficiency, or discomfort to occupants, possibly including injuries.

Hazardous Failure Failure conditions that would reduce the capability of the airplane or the ability of the crew to cope with adverse operating conditions to the extent that there would be:

- A large reduction in safety margins or functional capabilities
- Physical distress or higher workload such that the flightcrew cannot be relied upon to perform their tasks accurately or completely
- Serious or fatal injury to a relatively small number of the occupants

Catastrophic Failure Failure conditions that would prevent the continued safe flight and landing of the airplane.

Probable Failure Probable failure conditions are those anticipated to occur one or more times during the entire operational life of each airplane. Probable failure conditions are those having a probability on the order of 1×10^{-5} or greater. Minor failure conditions may be probable.

Improbable Failure Improbable failure conditions are divided into two categories as follows:

- **Remote Failure** Unlikely to occur to each airplane during its total life but may occur several times when considering the total operational life of a number of airplanes of the same type. Improbable (remote) failure conditions are those having a probability on the order of 1×10^{-5} or less, but greater than on the order of 1×10^{-7}. Major failure conditions must be no more frequent than improbable (remote).
- **Extremely Remote Failure** Unlikely to occur when considering the total operational life of all airplanes of the same type, but nevertheless has to be considered as being possible. Improbable (extremely remote) failure conditions are those having a probability of on the order of 1×10^{-7} or less, but greater than on the order of 1×10^{-9}. Hazardous failure conditions must be no more frequent than improbable (extremely remote).

Extremely Improbable Failure Extremely improbable failure conditions are those so unlikely that they are not anticipated to occur during the entire operational life of all airplanes of one type, and have a probability on the order of 1×10^{-9} or less. Catastrophic failure conditions must be shown to be extremely improbable.

Flag Operation means any scheduled operation conducted by any person operating any airplane described below at the locations described:

- Airplanes:
  - Turbojet-powered airplanes
  - Airplanes having a passenger-seat configuration of more than nine passenger seats, excluding each crew member seat
  - Airplanes having a payload capacity of more than 7,500 pounds

- Locations:
  - Between any point within the State of Alaska or the State of Hawaii or any territory or possession of the United States and any point outside the State of Alaska or the State of Hawaii or any territory or possession of the United States, respectively
  - Between any point within the 48 contiguous States of the United States or the District of Columbia and any point outside the 48 contiguous States of the United States and the District of Columbia
  - Between any point outside the United States and another point outside the United States.

Foreign Air Carrier is any person not a citizen of the United States who undertakes, directly or indirectly, or by lease or other arrangement to engage in foreign air transportation.

Foreign Air Commerce means the transportation of passengers or property by aircraft for compensation, the transportation of mail by aircraft, or the operation of aircraft in furthering a business or vocation, between a place in the United States and a place outside the United States when any part of the transportation or operation is by aircraft.

Foreign Air Transportation is the carriage by aircraft of persons or property as a common carrier for compensation or hire or the carriage of mail by aircraft in commerce between a place in the United States and any place outside the United States when any part of the transportation is by aircraft.

Intrastate Air Carrier means a citizen of the United States undertaking by any means to provide only intrastate air transportation.

Kind of Operation means one of the various operations a certificate holder is authorized to conduct, as specified in its Operations Specifications, i.e., domestic, flag, supplemental, commuter, or on-demand operations.

Noncommon Carriage means an aircraft operation for compensation or hire that does not involve a holding out to others.

On-Demand Operation means any operation for compensation or hire that is one of the following:

- Passenger-carrying operations in which the departure time, departure location, and arrival location are specifically negotiated with the customer or the customer's representative that are any of the following types of operations:
  - Common carriage operations conducted with airplanes, including turbojet-powered airplanes, having a passenger-seat configuration of 30 seats or fewer, excluding each crewmember seat, and a payload capacity of 7,500 pounds or less, except that operations using a specific airplane that is also used in domestic or flag operations and that is so listed in the Operations Specifications as required by FAR 119.49(a)(4) for those operations are considered supplemental operations
  - Noncommon or private carriage operations conducted with airplanes having a passenger-seat configuration of less than 20 seats, excluding each crewmember seat, or a payload capacity of less than 6,000 pounds
  - Any rotorcraft operation

- Scheduled passenger-carrying operations conducted with one of the following types of aircraft with a frequency of operations of less than five round trips per week on at least one route between two or more points according to the published flight schedules:
  - Airplanes, other than turbojet-powered airplanes, having a maximum passenger-seat configuration of nine seats or less, excluding each crewmember seat, and a maximum payload capacity of 7,500 pounds or less
  - Rotorcraft
- All-cargo operations conducted with airplanes having a payload capacity of 7,500 pounds or less, or with rotorcraft

Passenger-Carrying Operation means any aircraft operation carrying any person, unless the only persons on the aircraft are those identified in FAR 121.583(a) or 135.85. An aircraft used in a passenger-carrying operation may also carry cargo or mail in addition to passengers.

Principal Base of Operations means the primary operating location of a certificate holder as established by the certificate holder.

Private Carriage[1] does not involve offering or "holding out" to the public any offer of transportation services. It is rather:

- Carriage of an operator's employees
- Carriage of participating members of a club
- Carriage of persons or property incidental to the operator's primary business

Private carriage may also be the carriage of persons or property for compensation or hire under a contractual business arrangement which did not result from the operator's offering or holding out of service. It is usually for one or several selected customers, generally on a long-term basis. The number of contracts must not be too great, otherwise it implies a willingness to make a contract with anybody. A carrier operating pursuant to 18 to 24 contracts has been held to be a common carrier because it held itself out to serve the public generally to the extent of its facilities. Private carriage has been found in cases where three contracts have been the sole basis of the operator's business. Special adaptation of the transportation service

[1]See also FAA Advisory Circular, AC120-12A, Private Carriage versus Common Carriage of Persons or Property.

to the individual needs of shippers is a factor tending to establish private carriage but is not necessarily conclusive.

Private carriers for hire are sometimes called "contract carriers." The term is borrowed from the Interstate Commerce Act but is legally inaccurate when used in connection with the Federal Aviation Act.

Scheduled Operation means any common carriage passenger-carrying operation for compensation or hire conducted by an air carrier or commercial operator for which the certificate holder or its representative offers in advance the departure location, departure time, and arrival location. It does not include any operation that is a charter operation.

Supplemental Operation means any common carriage operation for compensation or hire conducted with any airplane described below for the types of operation indicated:

- Airplanes:
  - Airplanes having a passenger-seat configuration of more than 30 seats, excluding each crewmember seat
  - Airplanes having a payload capacity of more than 7,500 pounds
  - Each airplane having a passenger-seat configuration of more than nine seats and less than 31 seats, excluding each crewmember seat and any turbojet-powered airplane that is also used in domestic or flag operations and that is so listed in the Operations Specifications as required by FAR 119.49(a)(4) for those operations

- Types of operation:
  - Operations for which the departure time, departure location, and arrival location are specifically negotiated with the customer or the customer's representative
  - All-cargo operations

Interstate Air Commerce means the transportation of passengers or property by aircraft for compensation, the transportation of mail by aircraft, or the operation of aircraft in furthering a business or vocation:

- Between a place in:
  - A State, territory, or possession of the United States and a place in the District of Columbia or another State, territory, or possession of the United States
  - A state and another place in the same state through the airspace over a place outside the state
  - The District of Columbia and another place in the District of Columbia or

    - A territory or possession of the United States and another place in the same territory or possession and
- When any part of the transportation or operation is by aircraft

Interstate Air Transportation means the transportation of passengers or property by aircraft as a common carrier for compensation, or the transportation of mail by aircraft:

- Between a place in:
    - A state, territory, or possession of the United States and a place in the District of Columbia or another state, territory, or possession of the United States
    - Hawaii and another place in Hawaii through the airspace over a place outside Hawaii
    - The District of Columbia and another place in the District of Columbia
    - A territory or possession of the United States and another place in the same territory or possession
    - When any part of the transportation is by aircraft

Person as used in Sec. 314 of the Federal Aviation Act, is defined in Sec. 101(32) of the act and means any individual, firm, copartnership, corporation, company, association, joint stock association, or body politic; and includes any trustee, receiver, assignee, or other similar representative thereof.

Products are identified by class:

- Class I products are defined as a complete aircraft, aircraft engine, or propeller.
- Class II products are a major component of an aircraft, aircraft engine, or propeller, the failure of which would jeopardize the safety of the aircraft, engine, or propeller.
- Class III products are any part or component that is not a Class I or Class II product and includes standard parts, that is, those designated as AN, NAS, SAE, and so on.

Wet Lease means any leasing arrangement whereby a person agrees to provide an entire aircraft and at least one crewmember. A wet lease does not include a code-sharing arrangement.

When Common Carriage Is Not Involved or Operations Do Not Involve Common Carriage means any of the following:

- Noncommon carriage

- Operations in which persons or cargo are transported without compensation or hire
- Operations not involving the transportation of persons or cargo
- Private carriage

Index

Note: **boldface** numbers indicate illustrations; *italic* t indicates a table.

About the Author

Jack Hessburg has more than 40 years of experience in aviation, both in air carrier maintenance and flight operations with both airlines and the Boeing Company. He has worked for the U.S. and foreign scheduled and nonscheduled airlines in a variety of assignments. These have included positions in aviation, such as Director of Maintenance Training, Director of Flight Crew Training, maintenance instructor, and aircraft mechanical systems engineer. He taught in both the degree and mechanic programs at Park College, and during his 25 years at Boeing, he taught an after-work-hours course in air carrier maintenance and operation to employees. Jack is both a mechanical engineer and a licensed Airframe and Powerplant Mechanic.

He joined the Boeing Company in 1973. His first position was as flight operations engineer, supporting a number of United States, European, African, and Middle East operators. In 1980, he joined Boeing Field Service, providing on-site maintenance and engineering support to such airlines as United, Air Cal, World, Trans America, and Air Canada.

Jack returned to a factory assignment in 1986 as 747/767 airline support manager. In 1991 he was appointed Chief Mechanic for the Boeing Commercial Airplane Group and was the first person in aviation history to hold such a post. In this capacity he represented the interests of the entire maintenance community in the 777 airplane design. He brought the air carrier mechanic's and maintenance engineer's perspective to the design table, notably in the design of the 777 computer-based maintenance system.

Jack's efforts on behalf of maintenance features in the 777 design have been praised by airline customers and, in 1994, won him the Joe Chase Award from the Flight Safety Foundation and the Professional Aviation Maintenance Association. He was a recipient of the 1995 Commercial Aviation Engineering/Technical Achievement award from the American Institute of Aeronautics and Astronautics. In 1997 Jack received an Industrial Fellowship from the University of Exeter for his contributions to the field of systems maintainability. In 1999, he received the Lifetime Achievement Award in Aviation MRO from *Overhaul & Maintenance* magazine and an Honorary Doctor of Science from the College of Aeronautics.

Jack retired in January 1999. He continues to be active in aviation as a consultant, writer, and teacher.

DISK WARRANTY

This software is protected by both United States copyright law and international copyright treaty provision. You must treat this software just like a book, except that you may copy it into a computer in order to be used and you may make archival copies of the software for the sole purpose of backing up our software and protecting your investment from loss.

By saying "just like a book," McGraw-Hill means, for example, that this software may be used by any number of people and may be freely moved from one computer location to another, so long as there is no possibility of its being used at one location or on one computer while it also is being used at another. Just as a book cannot be read by two different people in two different places at the same time, neither can the software be used by two different people in two different places at the same time (unless, of course, McGraw-Hill's copyright is being violated).

LIMITED WARRANTY

McGraw-Hill takes great care to provide you with top-quality software, thoroughly checked to prevent virus infections. McGraw-Hill warrants the physical diskette(s) contained herein to be free of defects in materials and workmanship for a period of sixty days from the purchase date. If McGraw-Hill receives written notification within the warranty period of defects in materials or workmanship, and such notification is determined by McGraw-Hill to be correct, McGraw-Hill will replace the defective diskette(s). Send requests to:

McGraw-Hill
Customer Services
P.O. Box 545
Blacklick, OH 43004-0545

The entire and exclusive liability and remedy for breach of this Limited Warranty shall be limited to replacement of defective diskette(s) and shall not include or extend to any claim for or right to cover any other damages, including but not limited to, loss of profit, data, or use of the software, or special, incidental, or consequential damages or other similar claims, even if McGraw-Hill has been specifically advised of the possibility of such damages. In no event will McGraw-Hill's liability for any damages to you or any other person ever exceed the lower of suggested list price or actual price paid for the license to use the software, regardless of any form of the claim.

McGRAW-HILL SPECIFICALLY DISCLAIMS ALL OTHER WARRANTIES, EXPRESS OR IMPLIED, INCLUDING, BUT NOT LIMITED TO, ANY IMPLIED WARRANTY OF MERCHANTABILITY OR FITNESS FOR A PARTICULAR PURPOSE.

Specifically, McGraw-Hill makes no representation or warranty that the software is fit for any particular purpose and any implied warranty of merchantability is limited to the sixty-day duration of the Limited Warranty covering the physical diskette(s) only (and not the software) and is otherwise expressly and specifically disclaimed.

This limited warranty gives you specific legal rights; you may have others which may vary from state to state. Some states do not allow the exclusion of incidental or consequential damages, or the limitation on how long an implied warranty lasts, so some of the above may not apply to you.